Ethical Challenges in Genomics Research

Paula Boddington

Ethical Challenges in Genomics Research

A Guide to Understanding Ethics in Context

Springer

Dr. Paula Boddington
Somerville College
Oxford University
Woodstock Road
Oxford OX2 6HD, UK
paula.boddington@some.ox.ac.uk

ISBN 978-3-642-23698-3 e-ISBN 978-3-642-23699-0
DOI 10.1007/978-3-642-23699-0
Springer Heidelberg Dordrecht London New York

Library of Congress Control Number: 2011943164

Printed on acid-free paper

Springer is part of Springer Science+Business Media (www.springer.com)

This book is dedicated to the dear friend of my schooldays with whom I happily shared many science lessons, Damyanti Patel.

Acknowledgements

My first thanks must go to my scientific colleagues from the Procardis consortium, based at Oxford, Stockholm, and Milan, in particular to Anna Helgadottir, and also to Hugh Watkins, John Peden, Anuj Goel, Theo Kyriakou, Martin Farrell, and Robert Clarke. Many other colleagues, too many to name, have provided inspiration, comradeship, and good cheer: prominent in the lengthy list are Naomi Hawkins, Jantina de Vries, Maggie Gregory, Susan Hogben, Ulla Räisanen, and Paul Atkinson. Thanks also for comments, suggestions and guidance from Jane Gitschier and from Philip Zimbardo.

The work upon which this book is based was carried out whilst employed by the Procardis programme, which is funded by the Sixth Framework Programme of the European Union, project number LSHM-CT-2007-037273.

Contents

Chapter 1
The "Ethics Job"

Abstract This chapter introduces the aims of the book. The book's origins lie in a project examining ethical issues in genomics research, as part of a consortium researching into cardiovascular disease. The challenges of working as an "in-house" ethicist on such a project are discussed, as are the importance of debate and dialogue in ethics as well as inter-disciplinary communication. The book aims to help foster such dialogue and to show how philosophical analysis and debate in ethics can help to address emerging ethical challenges. Some of the different ways in which philosophy can contribute to practical ethical issues are illustrated. These include close attention to argument; to the use of key concepts; and to the use of the examples used to advance arguments. The interrelation of abstract theory and experience of concrete situations is also discussed.

1.1 The Role of the "In-house" Ethicist and How This Book Came About

I am a philosopher. In October 2007, I took up a post in the Medical Sciences Division of Oxford University and began work on the "ethics" branch of a scientific research project investigating genetic factors underlying cardiovascular disease. The Procardis consortium includes members from various European countries, including the UK, Italy, Sweden, Germany, and France, and is funded by the European Union. But why is research that aims to help humanity subjected to ethical inquiry? Why do they want someone to "do" ethics then? What was this job exactly?

When I started my post, I did what many other people do in such a position; I read through the large black file about the project that was handed to me, looked back over the job description, and started to wonder what exactly I was meant to be doing. Writing this at the end of my 3-year post, I am still thinking about what exactly such a post of being an "in-house" ethicist required. There are also various tensions inherent in such a post, tensions which could easily give rise to a great deal

DOI 10.1007/978-3-642-23699-0_1,

of difficulty. This book is a product of thinking through these difficulties, and of my efforts to produce something that might contribute to thinking through the complex issues that I have encountered, issues of both methodology and substance.

One of the features of my particular position in the project was that it was, quite frankly, rather loosely demarcated. Some colleagues who worked in other broadly similar situations have had a more defined role as a member of a team. For example, these colleagues have been involved with obtaining approval for research projects from the relevant research ethics committees; with drafting policy for the conduct of a research project, taking into account particular ethical issues; with looking at how best to obtain adequately informed consent to research from people in difficult circumstances, such as when their child is ill, across language barriers, or where research recruits have little or no prior knowledge of the scientific study of genetics. Some of these colleagues were frequently called over to the scientists' building to help out on one thing or another.

In contrast, I had no such concrete role. The subjects in the Procardis project were all recruited long before I joined; when they joined, they had read and signed consent forms, and ethics approval for the project and for the recruitment of participants had of course been given. This in fact gave rise to one of the tensions of such a post. Suppose I looked at what had already been done and found it ethically problematic in some respect? In any case, much of this was someone else's job to check – the job of the various research ethics committees that approved the project and the collection of data in several European countries. In any case, I felt some discomfort at the idea that my role was to tell the scientists that they might be doing something wrong. My job was surely not to act as a kind of soft police force. Besides which my only tools were words: I had no regulatory or institutional power.

It also filled me with great discomfort to think that the role of an "in-house" ethics person might be, even partially, to act as "window dressing". There is now a great deal of emphasis on ethics in genetics, for reasons I'll explain in detail in the next chapter. The simple fact then is that *doing ethics* has come to "look good". Of course, many things which "look good" are good, but not all, and pressures to look good can lead to an unwarranted concern with appearances. Moreover, I consider that the goal-oriented nature of much research funding carries with it a danger that doing good work can be reduced to meeting targets and scoring highly enough in the right sort of categories. The very thought that the "ethics" person might be symbolised as a kind of talisman to indicate lip service to ethical issues was a spur to make sure that I thought carefully about the nature of my role and how to be most effective.

An ethics post attached to a particular consortium gave me a delicate balancing act. On the one hand, I was part of this team: to that extent then, I shared their goals and hoped for their scientific successes; as I got to know the other team members, I liked them. I wasn't there to "judge" them. But I should surely, as someone concerned with ethical issues of the research, also remain independent, and retain the capacity and the freedom to offer critiques or challenges if I thought it necessary. In many ways, the problem of how to tread this delicate balance between involvement and detachment is one well known from other areas. For instance, the ethnographer who lives intimately among a group of people in order to document their lives and activities must identify with these people to a sufficient degree to be able to observe and understand; but then would understanding and description be compromised by over-identification, by "going native"? What of the ethnographer who uses the material to publish a work critical of the group of people amongst whom he or she has lived? (Hammersley and Atkinson 1986)

How could I steer a course through such difficulties? My job description required me to run a series of seminars and workshops about ethics across the consortium, although there was little demand for them from the extremely busy scientists. Much of my time was in fact taken up with writing research papers, often in collaboration with colleagues who are social scientists and lawyers and who were working in related areas of the ethics and governance of genetics and of biomedical research more generally. In engaging with the scientists however, I came to consider that there was a strong general need on all sides for further mutual understanding and dialogue between those with different disciplinary backgrounds. There is great scope for building the capacity to think through complex issues in ethics; this would be the answer to my dilemma of how to navigate the demands of my post – by attempting to contribute to the task of interdisciplinary understanding, and, instead of "telling scientists what to do", to build capacity for individuals to think through complex ethical issues for themselves. The book is a development of that work.

1.2 Research in Genomics

Before going on to describe in more detail what thinking about ethics involves, and what this book in particular will involve, it will be helpful to fill in a little about what research in genomics involves.

1.2.1 Genetics and Genomics

Firstly, it is useful to explain the distinction between genetics and genomics. *Genetics* generally refers to a focus on specific genes, for example, specific genes that might be involved in certain diseases such as cystic fibrosis, where one or a few genes are involved. But individuals have about 20–25,000 genes, and well as a great deal more genetic material arranged on their 23 paired chromosomes. *Genomics* refers to the study of large amounts of genetic information, including spanning the whole of the human genome. This book focuses on developments in genomics research, although there are overlaps between ethical issues in genetics and in genomics.

1.2.2 A Genomics Research Project: The Procardis Consortium

The Procardis project can be used as one example of genomics research. The focus of the work is on cardiovascular disease, in particular, looking at people who have suffered from early onset myocardial infarction (heart attack) and comparing them to healthy "controls" (people of a similar age and the same gender but who have not suffered a heart attack) in an attempt to find any genetic differences between these two groups. Because cardiovascular disease is a very common, complex condition, there are a large range of different factors involved in its genesis, including what might be broadly termed environmental factors such as diet, and many complex interacting biological factors, such as blood pressure, blood lipid levels, and a great deal more. There are very likely to be also a large range of genetic factors involved, and it is also likely that each one of these may contribute only a small amount to the overall picture. Because cardiovascular disease is complex, and because each genetic signal is likely to be weak, this also means that many thousands of recruits are needed to get a sufficiently large sample to obtain significant results. The genetic signals that are found to be associated with disease then need to be further investigated to try to understand the underlying biological mechanisms that cause disease or protect against it.

For example, Procardis have investigated links of coronary artery disease with blood levels of C reactive protein (Elliott et al. 2009), blood plasma levels of homocysteine which indicate novel metabolic pathways in need of further investigation (Mälarstig et al. 2009) and genetic associations with the blood lipid Lp(a) lipoprotein, which strongly suggest that genetic variations in this lipid are causally implicated in coronary artery disease (Clarke et al. 2009).

1.2.3 Varieties of Genomics Research

There are a great deal of other genomics research projects underway, many investigating common complex diseases, others investigating less common diseases, many undertaken as part of wider investigations into health. Some involve looking at different reactions individuals have to various drugs. Some projects are part of large biobanks, where large amounts of medical information about large numbers of people are stored for research use over many years. Some involve fairly small numbers of people, for example, individuals suffering from unusual diseases; some involve many thousands, or hundreds of thousands of individuals. Some are located at a particular institute; others are national or often international consortia or large collaborations between many different consortia. Genomics research aims to further our understanding of the biology of health and disease, through understanding the contribution of genetic information to disease, the interaction of genetics and environment, broadly understood, and the interaction of different genes. Hopes for this research are a better understanding of the biology of disease causation, with eventual hopes for better treatment and diagnosis (Heard et al. 2010).

1.3 Ethical Debate in Genomics

There are in fact a great many questions about ethics and about the regulation of research presented by scientific and technological developments in genetics and genomics. There are questions raised by the uses to which such research findings may be put. There are questions raised by the immense power and detail now seen in the genetic information that it is possible to obtain about individuals, about populations, and about human beings in general. There are questions about protecting the privacy and other interests of research participants, and questions about what might or might not be owed to them. There are questions about how data from genomics research might be shared with other scientists doing similar work.

Besides this issue of what ethical questions there are, there is also the issue of how to go about addressing any such questions. I started this chapter with the statement "I am a philosopher". One reason for this is that it forms a firm statement of my disciplinary basis; although the ethics work I have been doing might broadly be described by many as falling within the realm of bioethics, bioethics itself typically inhabits a far narrower realm than philosophy. I strongly believe that it is crucial to keep the wider focus of philosophy. There are a whole host of considerations from diverse branches in philosophy that can be brought to bear upon issues in ethics. For example, issues in the philosophy of science can be very pertinent in considering questions in genetics and in medicine: for instance, how we define and think about health and disease. In my own case, work I have done within philosophy of mind, on self-deception, has led me to consider the problems of irrationality that mar much of our thinking and action, especially where ethics is concerned, and to ponder the crucial importance of motivation in ethical reasoning and action. Questions such as these will be raised throughout this book.

Philosophy concerns itself, centrally and self-consciously, with reasoning, evidence, and argument. Of course, all disciplines are concerned with these questions, and are concerned to think critically about their subject matter, but philosophy sets out to think critically about critical thinking – an important aspect of doing philosophy concerns the analysis and dissection of positions and views, in illuminating problems as much as solving them. In teaching philosophy, a large part of what one is doing is teaching the capacity to think things through for oneself. This can be useful for many reasons, including enhancing communication and understanding between those with different disciplinary backgrounds, and for assisting with enriching democratic participation in the public debates and policy discussions on these complex issues. More will be said to emphasise the importance of clear and open debate in medical research later.

Much work in ethics, especially in bioethics, is geared towards providing an answer to ethical questions. Is it acceptable to use human–animal chimeras in research? Should pre-natal selection for sex be permitted? Should participants in biobanks and in genomics research be given their personal results? The way I have always preferred to work has a different focus: on looking at the details of how debates are constructed. This involves strategies such as attempting to explain arguments and positions; pointing out implications of views, their consequences, and possible inconsistencies; assisting with clarification of the concepts involved; and attempting to show the basis behind opposing views. Such an approach is in any case a good groundwork to reaching a conclusion on a question, or at least an acceptable compromise. But additionally, it is also a good approach for assisting others to comprehend the debates and hence to join in. It is the philosophical equivalent of "showing your workings out" in maths; displaying these will best enable others to engage in the argument, and to see where the strengths, weaknesses, and assumptions of one's position lie.

Wittgenstein remarked that "in philosophy, the winner of the race is the one who can run most slowly. Or the one that gets there last" (Wittgenstein 1980). This may sound like a council of folly in a world where the race is on to publish before a rival research team pips you to the post, or before your funders require an account of what you have been doing. But of course, those engaged in such speedy work also know fully well that it has to be done properly and with full rigour. (And I will have more to say about genomics and speed later.)

In addressing questions of ethics, as in many other areas of life, we need to take a broad perspective. We need to take into account the viewpoints of all those who are involved and who might be affected in some way. We also need to gather a great deal of evidence, and often this means drawing on the experience and knowledge of others. Some of the things we need to address in debates in ethics are facts – or at least what one might call "expert opinion". What resources do you need to conduct a certain piece of research? Is it possible to ensure that nobody will be able to trace an anonymised DNA sample back to the individual? Hence again we see the great importance of debate and open discussion with others from a range of backgrounds, opinion, agendas, and experience.

This is reinforced in my experience by the realisation that it is often hard to know in advance what pieces of information might turn out to be relevant to a debate, so that it is hard then to go to "the experts" with a set list of inquiries. There needs to be more open dialogue. It is also often the case that wires are crossed in communication because different people are using the same terms but in different ways (we will look at some examples of this later in the book). And one is more likely to get useful dialogue the more we all understand each other. This all underscores the need for deepening understanding between different concerned groups and individuals, and fostering the capacity of others to participate in ethical discussion.

This is what this book is all about.

1.4 Putting Philosophy to Work in Ethics and Genomics

What value might philosophy be in all this? Philosophy often begins by addressing the sceptic, the critical questioner who looks for weaknesses in our arguments and asks probing questions about the hidden assumptions behind our assertions. It might be useful to provide some illustrations of the sort of work in philosophy that may help to clarify and assist debates in ethics.

1.4.1 Careful Attention to Steps in Arguments

Susan Hogben, a discourse analyst, and I together examined policy statements concerning genetic testing of children, concentrating on position papers which included the reasoning and arguments leading up to the statements of policy. Discourse analysts pay particular careful attention to language and how it is used; in this, they have a great deal in common with philosophers, and the origins of the discipline have significant overlaps with philosophy. Throughout this book, I will be examining some important examples of how close attention to language can greatly influence debates in ethics. I will also encourage readers to look carefully at how examples are chosen and used in debate.

We looked carefully at how examples of genetic conditions are picked in arguments supporting policy statements about childhood genetic testing (Boddington and Hogben 2006). Whenever examples are chosen, one naturally has in mind that these are in some way "typical" and that broad general conclusions applicable to other cases can be drawn from them. In fact, in reading one is often tempted to skip over examples quickly on the assumption that they do no more than provide a simple clarification of a general point. But not so fast! We concluded that the choice of examples, together with how they were presented, could be shown to have serious impact on the conclusions drawn; with a different choice of examples, different policy conclusions might have been reached.

For example, the choice of Huntington's disease was consistently used, by different groups, as an exemplar of a late onset genetic condition. There may be historical reasons for this, because it was one of the first such conditions for which testing was available. But this disease has several distinct features that make it a poor model for other conditions. I suspect that it is often used as an example because it is such a vivid and clear illustration of a genetic disease. But this in itself precisely helps to make it a poor exemplar. It is a degenerative, neurological condition, affecting cognitive capacity; it is fatal; there is no treatment currently available; its transmission is dominant which means that on average half of any children of affected people will also be affected; it is fully penetrant, meaning that if one has the gene in question, one will certainly develop the disease; moreover, often it is possible to predict the rough age of onset by testing for the exact mutation a person carries. All this makes Huntington's disease relevantly different from many other late onset genetic conditions which may be expressed to varying degrees in different family members, which affect different organ systems, which may not be fatal, or for which some treatment or palliation may be available. We concluded through this close analysis of the documents that drawing on a wider and more representative range of examples might well have led the policy makers to different conclusions.

This leads us to one of my favourite aphorisms on how to conduct philosophy: "A major cause of philosophical disease: an unbalanced diet. One nourishes one's thinking with only one kind of example" (Wittgenstein 1974). It is a warning that reinforces the need for discussion with others, and for casting one's own mental net widely, for this can be a good source of enriching one's store of examples. Critically, it underlines the need for dialogue between disciplines and those with specialist knowledge in particular areas – for example, here, some understanding of Huntington's disease and of the range of genetic conditions and their characteristics is needed in order to appreciate the ways in which Huntington's differs from other possible examples. When thinking through ethical issues, the need to consider all points of view is never more pressing. This should be a familiar lesson for scientists to take on board, since different diseases have very different underlying biology, and so choice of disease to study can make a large impact upon how research is conducted, and upon what general lessons might be learned about genomics.

1.4.2 Matching up Theory with Experience

Other work looking at social and ethical aspects of genetics has shown the importance of grounding theoretical work in an appreciation of the lived experiences of those affected. For example, I worked with Maggie Gregory, a sociologist who interviewed many people from families affected with various genetic conditions, including the blood disorder haemophilia. She was interested in how families communicate about health and illness. This topic raises a great deal of important ethical questions, and we will look at some of them later in the book. In particular, there is a large philosophical and legal literature which deals with the question of who has a right to genetic information. This question can pose particular difficulties in the context of genetics, since information about one person can potentially be of relevance to others to whom one is genetically related. For example, if a woman gives birth to a child with a certain condition, and therefore discovers that she herself is a carrier of the condition, this often means that her sister may likewise be a carrier. Should she tell her sister? Does her sister have a "right" to know? (Bell and Bennett 2001)

The philosophical and legal literature on this has paid considerable attention to the question of the "right" to genetic knowledge, and conversely, the right to refuse such information. The literature is geared towards answering the question of whether or not someone should be told – to questions of rights and to questions of the infringement or enhancement of autonomy. But Maggie's work in families with haemophilia found that these concerns did not even seem to be on the agenda.

Nobody in her sample seemed to raise the question of "should I tell or should I not"? They were concerned with issues such as when is the best time to tell; with who can handle the information; with how to tell; who is the best person to convey the news. Work in philosophy had by and large overlooked these issues (Boddington and Gregory 2008).

Furthermore, ethical debates about autonomy and genetic information have been concerned with issues such as how such information might enhance rational decision making and might therefore enhance freedom and choice. The range of issues that preoccupied those to whom Maggie spoke however seemed broader than this: genetic information was not seen only to be about making decisions or being maximally informed, but fitted into to a richer, wider, and more emotional world. For example, one woman had been brought up by her mother together with, from a very early age, a loving step-father, and could not remember her genetic father who died when she was small. She showed little interest in following up her status as a carrier of a condition she had inherited from her genetic father: failing to equip herself with such information might be seen to lessen her ability to act on the greatest range of relevant information. But she saw this as a sort of disloyalty to her step-father, who had brought her up from infancy, and because of this, she preferred not to "go there". The integrity of her current family unit was her prime concern, rather than a more narrowly focussed notion of autonomy.

Work such as this led us to question the primacy of autonomy as a key concept in the philosophical and legal debates about genetic information and to reconsider whether or not talk of "rights" to information captured all that we need to consider. The lessons from this example are particularly pertinent to this book too as we shall see how prominent a place autonomy had had in medical and research ethics, but how genetics and genomics research in particular produce challenges to the conception and role of autonomy in ethical practice.

These considerations, drawn from analysis of empirical work, also cohered with other, more theoretical reasons I had for questioning the primacy of autonomy in this and other ethical debates. There is a two-way flow between what theories we adhere to and how we see the world. This is one of the reasons we need multiple voices and multiple eyes, and again, we shall return to this point in more detail throughout the book.

This is an example of how close attention to what actually concerns people may enrich the conceptual framework we use to make sense of our world, our dilemmas, and to pose questions and find solutions. It is an example of the useful process of going forwards and backwards between our abstract conceptual framework and concrete experiences. We should also remember, however, the lesson we learned from considering choice of example: the sample here was of families experiencing haemophilia, and moreover, in most if not all of the families, the haemophilia was relatively mild and relatively well controlled. There was also virtually universal high praise for the medical facility where this group was receiving treatment. In other situations, and maybe where other sorts of conditions were concerned, questions of "rights" to information and questions of autonomy and the quality of decision-making may come more to the fore.

Lessons so far: think carefully about our choice of examples and about how we draw general principles from them; think carefully too about how our concepts and the theoretical framework for analysing problems fit with real life and experience. We need to be clear about what we mean by certain concepts we use, and what their implications are, and whether this is leading us astray or into erroneous conclusions and problematic value judgements. In Chap. 3 we discuss in more detail the importance of attention to the use of concepts in general and to certain key concepts in ethics in particular. Later in the book, we will look in particular at the concepts of autonomy, of respect for persons, and of responsibility.

Often texts in bioethics start by looking at general principles, giving a broad brush introduction to the subject and attempting to provide a synoptic overview. But this can give a false impression of security, an overoptimistic sense that that's all that's needed. It can lead some to go rushing off armed with a hastily gathered bundle of principles or key concepts, causing mayhem. Starting off with general principles also might not be the best approach if one wishes to critique some of the mainstream methodological approaches to bioethics.

I have discussed how theory and practice may interact, and how our assumptions may blind us to what is there, and how questioning these can lead to different answers. There are many different ways of approaching the same subject matter, and these may have their own strengths and weaknesses. In biology, some approaches look at the entire ecosystem; some at the systems of whole organisms; and some at organ systems within organisms, at cellular, subcellular, or molecular levels. Often it is necessary to move freely between one level and another. Often it is important to triangulate, to approach from different angles and apply with different tools and concepts. Different approaches can work well in different situations, or in complement with each other.

1.4.3 Developing Moral Thought

I wish to tread the line between encouraging and enabling the reader to think carefully about ethical issues, whilst avoiding the impression that a brief introduction is all that is needed. Of course, we all think about ethics in our everyday life, and it would be grossly wrong to give the impression that it is only the province of professionals or that "we" do it better than "amateurs". I hope my discussion of the vital need for dialogue makes this clear. Nonetheless, I think that there can be much to be gained from closer and rigorous attention to arguments and concepts used in thinking about ethics, else I would not be writing this book. There will also be much to say about difficulties with the operation of individual moral codes and individual consciences within a social, and especially within an institutional, context. These considerations will give an especial impetus to the need to think very rigorously about ethical judgement, ethical motivation, how it might go awry, and how it might be made more robust.

Phronesis

Phronesis is a Greek word often translated as "practical wisdom", which refers to the capacity to make mature, skillful moral judgements. Aristotle discussed *phronesis* at length in his major work on ethics, *The Nicomachean Ethics* (Aristotle 1999). It is a complex concept and there are some differences of opinion about how exactly it should be understood, but it is a useful concept to think of here in explaining the project of this book.

Aristotle saw that it was necessary to pay particular attention to how people could develop a growing capacity for moral judgements and actions, and how we could develop virtuous character traits. This required training in acting in the right way, as well as the development of appropriate feelings and of a capacity to make fine judgements needed in different occasions. Crucially, Aristotle assumed in writing his work on ethics that he was addressing an audience who already were on his side: who already had been "well brought up"; in other words, people who knew, at least broadly, how to act morally and, moreover, who wanted to act well. Nonetheless, philosophical discussion and reflection on moral questions was seen as of value in the development of *phronesis*, of practical wisdom, in acquiring the skills needed to make finely tuned moral judgements in difficult circumstances, and in developing the character traits that led one to have the right feelings and to choose the right action. Aristotle also started from a base of the commonly accepted morality of the time; but saw that philosophical reflection on morality might lead one to revise this. Aristotle also saw that we needed knowledge of particular circumstances in order to act well morally. A combination of action, feeling, and thought is required for morality (Aristotle 1999).

Interestingly for our work, Aristotle was also the world's first biologist, making extensive and systematic observations of the animals around him. These formed the basis for education in biology for many centuries.

The focus on philosophical methodology and techniques is also a way of treading a line between regarding ethics as the domain of the specialist and of dismissing the idea of "ethics" as an academic enterprise. The discussions above about the interweaving of philosophical argument with everyday thought and experience, together with the constant need to bring into play the perspectives of others, mean that there is room, and indeed need, for all. We are aiming at practical wisdom, at what we might, following Aristotle, call *phronesis*.

We all have skills or reasoning, argument and analysis, but they can be greatly honed, and skills that we have unconsciously can often benefit from being made more conscious. Likewise, assumptions and beliefs we hold can also often benefit from being made more explicit, and from further analysis. Moreover, increased knowledge of the philosophical literature and of traditions of thought and argument in ethics can provide a rich resource with which to expand our boundaries and the habitual ebbs and flows of our thought.

I also should therefore give a disclaimer. This book is intended to look in particular at contemporary ethical issues arising in genomics research ethics, and in doing so it will also survey many key issues in research ethics, as well as looking at some key concepts used more generally in ethics. But it does not set out to be a comprehensive overview of these areas. This is partly because there are so many of these issues; and it is partly because in any case with the development of science and technology, these issues are constantly changing and new problems are presenting themselves. It is also partly because I think a better way of showing interested readers how to enhance their capacity to engage in ethical debate is to look in detail at some problems than attempting to give a broad but more superficial sweep of all the issues. In looking at particular issues, useful techniques for approaching ethical questions will be explained and illustrated.

Some people have talked about DNA as a blueprint, a recipe, a handbook, as if it told all that was needed to build an organism. We know that is not true. A lot more is involved, including interactions with the environment within which the organism finds itself. Likewise, this book too is not enough, and philosophy is not enough – it is far from a blueprint, not even a recipe, just a few pointers. The philosopher of science David Resnik has also argued that there is no "blueprint" for science ethics education (Resnik 1998). You will need, at the very least, to interact with your environment, as it were, and bring to bear experience along with more theoretical strategies. Part of the reason why Aristotle's discussion of *phronesis* has left scholars wondering how exactly he meant us to achieve this is precisely because it is hard.

One thing we need to do in ethics is to take a broad view of the interests of all those involved, and one of the central questions in research ethics in general is that of balancing the interests of subjects with the interests of research and of researchers. This book aims to try to improve dialogue about ethics between different academic disciplines. Likewise, I hope that those who might be volunteering to take part in genomics research can also use it to enhance their ability to think through the ethical issues involved in research participation. In so far as such readers find the book helpful, it can therefore play a small role in addressing one of the big issues of research, and in particular of genomics research, the general problem of public understanding of and engagement with science and the related ethical questions. I hope that you will be willing and enthusiastic about participating in the remainder of this book.

1.5 Conclusions

Pay close attention to the wisdom of multiple disciplines: Philosophical analysis needs to walk hand in hand with research from other disciplines.

Be careful about the choice of examples used to illustrate points or from which conclusions are drawn. Think through how things might go with a different example.

Pay close attention to the stories people have to tell about their own experiences, and think how broader value concepts do, or do not, help to make sense of such stories.

Look carefully at how we understand and use key concepts: look for inconsistencies and gaps.

Ethical issues may be at the forefront of discussion; they may also be hidden.

1.6 Study Questions

1. Think about an ethical question that you have encountered. How would you describe the problems? What concepts do you draw upon? With whom might you usefully discuss this question – and what would different people with different expertise and experience have to offer?
2. What are the advantages of appointing someone to a scientific research team to look at ethical issues? What are the disadvantages? Would it be better for such a person to have a degree of independence? What kind of training might such a position require?
3. Many philosophers have grappled with the question of what is involved in practical wisdom (*phronesis*), with developing good ethical judgement. What are the characteristics of somebody who possesses such "practical wisdom"? Is this something anyone can develop?
4. A related question: Aristotle assumed that the readers of his work on ethics were motivated to act morally and had been "well brought up". What is needed for ethical education, in order to foster good moral development?
5. What ethical questions regarding research into genomics concern you?

1.7 Coming Up Next

A discussion of why there is so much attention to the question of ethics in genetics and genomics research.

References

Aristotle (1999) Nicomachean ethics (trans: Irwin T), 2nd edn. Hackett, Indianapolis

Bell D, Bennett B (2001) Genetic secrets and the family. Med Law Rev 9:130

Boddington P, Gregory M (2008) Communicating genetic information in the family: enriching the debate through the notion of integrity. Med Health Care Philos 11(4):445–454

Boddington P, Hogben S (2006) Working up policy: the use of specific disease exemplars in formulating general principles governing childhood genetic testing. Health Care Anal 14:1–13

Clarke R, Peden JF, Hopewell JC, Kyriakou T, Goel A, Heath SC, Parish S, Barlera S, Franzosi MG, Rust S, Bennett D, Silveira A, Malarstig A, Green F, Lathrop GM, Gigante B, Leander K, de Faire U, Seedorf U, Hamsten A, Collins R, Watkins H, Farrall M, Consortium P (2009) Genetic variants associated with Lp(a) lipoprotein level and coronary disease. N Engl J Med 361(26):2518–2528

Elliott P, Chambers JC, Zhang W, Clarke R, Hopewell JC, Peden JF, Erdmann J, Braund P, Engert JC, Bennett D, Coin L, Ashby D, Tzoulaki I, Brown IJ, Mt-Isa S, McCarthy MI, Peltonen L, Freimer NB, Farrall M, Ruokonen A, Hamsten A, Lim N, Froguel P, Waterworth DM, Vollenweider P, Waeber G, Jarvelin M-R, Mooser V, Scott J, Hall AS, Schunkert H, Anand SS, Collins R, Samani NJ, Watkins H, Kooner JS (2009) Genetic loci associated with c-reactive protein levels and risk of coronary heart disease. JAMA 302(1):37–48. doi:10.1001/jama.2009.954

Hammersley M, Atkinson P (1986) Ethnography: principles in practice, 2nd edn. Routledge, London

Heard E, Tishkoff S, Todd JA, Vidal M, Wagner GP, Wang J, Weigel D, Young R (2010) Ten years of genetics and genomics: what have we achieved and where are we heading? Nat Rev Genet 11(10):723–733

Mälarstig A, Buil A, Souto JC, Clarke R, Blanco-Vaca F, Fontcuberta J, Peden J, Andersen M, Silveira A, Barlera S, Seedorf U, Watkins H, Almasy L, Hamsten A, Soria JM (2009) Identification of ZNF366 and PTPRD as novel determinants of plasma homocysteine in a family-based genome-wide association study. Blood 114(7):1417–1422. doi:10.1182/blood-2009-04-215269

Resnik DB (1998) The ethics of science: an introduction. Routledge, London

Wittgenstein L (1974) Philosophical investigations (trans: Anscombe GEM). Basil Blackwell, Oxford

Wittgenstein L (1980) Culture and value (trans: Winch P). Basil Blackwell, Oxford

Chapter 2
Why Research Ethics? Why Ethics in Genetics? The Case of Genomics Research

Abstract This chapter gives a brief history of the ethical regulation of research, discussing the original aims of such regulation, including the need to protect research subjects, to foster researcher responsibility, and to curtail the abuse of power. In addition, the need for openness of information and debate is emphasised. A brief history of ethical attention to issues in genetics and genomics is also given together with an analysis of why genetics and genomics have received such ethical scrutiny. Eugenics, liberal eugenics, and genetic determinism, essentialism, and reductionism are all discussed. The more recent research on ethical, legal, and social implications of genetics is introduced, and problems of research on human subjects are further explored. The chapter ends with a discussion of some recent examples of genomics researchers facing ethical questions.

2.1 A Brief History of the Ethical Regulation of Research

The scope of our inquiry is ethical issues arising within genomics research. This inquiry lies at the intersection of different areas of ethical inquiry. It lies within the wider field of research ethics; within the field of questions that arise in relation to genetics and genomics; and within the field of medical ethics more widely. As a preliminary, we need to understand why there is at present so much emphasis on these ethical issues. A good place to start is by looking at the past, with an examination of the history of research ethics regulation and of the context within which codes of research ethics and mechanisms for the governance of research were first drawn up.

This is not simply history that tells a story about days gone by; it tells a story of a vital influence on present thinking about research ethics, genetics, and genomics. An appreciation of this history will be an invaluable foundation for understanding the particular challenges that are faced currently by research in genomics. It will also serve as a basis from which to consider the rationale behind the various regulations that are in place. Codes of research ethics are themselves subject to

P. Boddington, *Ethical Challenges in Genomics Research*,
DOI 10.1007/978-3-642-23699-0_2, © Springer-Verlag Berlin Heidelberg 2012

evolution and interpretation and reinterpretation over the years. Hence, we may end up with understandings and interpretations that diverge from the original intentions. Our brief foray into history will also provide a context for a discussion of more general questions about how to approach ethical issues in research.

As is well known and amply documented, historically the major international codes of research ethics originated in the pressing need to protect individuals from what are widely recognised as extremes of abuse at the hands of unscrupulous and unethical researchers and institutions. Notorious amongst the notorious in the annals of medical infamy are those abuses carried out under the Nazi regime and exposed in the Nuremberg trials of 1947, culminating in the Nuremberg Code of research ethics. Indeed, the history of the development of thought regarding research ethics contains numerous examples of ethically problematic practice which have individually and collectively helped to shape thought. Some of these will be discussed in more detail later in the book.

However, although the practice of formalised codes of research ethics is still widely thought of as having its inception at Nuremberg, guidelines and regulations are found even earlier than this. At the end of the nineteenth century in Prussia, criticisms including the collection of 600 cases of unethical research on human subjects by Albert Moll led to the issuing of a directive to hospitals and clinics requiring consent to research. In Germany, criticisms in the press and by politicians of unethical human experimentation had led to the issuing of ethical guidelines by the Reich government in 1931, which were not repealed by the Third Reich. We will consider in due course the implications of the fact that there existed ethical guidelines for research in the very milieu in which unsurpassed atrocities occurred.

Circular of the Reich Minister of the Interior: Guidelines for New Therapy and Human Experimentation, Germany, 1931

Distinguished "therapeutic" from "nontherapeutic" research.

Included the principles of beneficence and nonmaleficence.

Based upon an ideal of patient autonomy.

Outlined a new legal doctrine of informed consent: consent had to be clearly given and required appropriate information.

New therapy could be introduced without consent only if it was urgently required.

For nontherapeutic research, consent was mandated.

There should be written documentation of consent procedures.

A clear structure of responsibility for each clinical trial was required.

Experimentation on the dying was prohibited.

Publication of results must respect dignity and the patient's humanity (Vollmann and Winau 1996).

The Nazi medical experiments, many of which were carried out under the auspices of the notorious Joseph Megele, were conducted on Jews, Gypsies and others deemed "defective" in some way, as well as on twins. Subjects were exposed to horrendous treatments involving extreme suffering, both physical and mental, and death. These included experiments on hypothermia, as well as studies on the genetics of eye colour. Many of these experiments are considered to be very poor science, although historically there has been some dispute about their scientific merits (Weindling 1996); they certainly bore every sign of utter disregard for the welfare of the subjects.

It seems an overwhelmingly obvious moral imperative that individuals should be protected from the abuses such as those carried out in the name of medical research by doctors and others operating in the Third Reich. The regime under which these abuses occurred treated the subjects of medical experiments purely as means to an end. The dangers of clinical research for the subjects are those of the risks of physical harm, and also emotional, psychological, and social harm resulting from the experimental procedures. These include the administration of novel drugs and therapies including surgical procedures. Risks up to and including death are faced in such research. Research ethics regulation has thus attempted to minimise such risks to those participating.

There are many other examples of what is now considered unethical research which may have fallen short of the extreme abuses of the Nazi regime but which nonetheless are now seen to treat the subjects of research with less than adequate regard. Many studies which have become classic examples of abusive research bear the hallmarks of exploitation and of the abuse of power. These include, for instance, the Tuskegee syphilis study which recruited African–American men and the Willowbrook hepatitis study which studied handicapped children resident in an institution (Murphy 2004). (Tuskegee will be discussed in greater detail later in the book.)

It is no coincidence that many cases that have been seminal in shaping attitudes to research ethics involved experimental subjects in weak social positions: African–Americans in the Southern United States in the earlier part of the twentieth century prior to the civil rights movement; residents of institutions such as prisons; the intellectually handicapped; children; and members of the armed services. For those imprisoned by the Nazi regime and subject to forced experiments, the power imbalance is obvious; in many other cases, low social status, lack of education whether general or lack of particular knowledge, poverty, age, or infirmity have created particular power imbalances; and in all cases, the power and social standing of the medical profession, as well as the prestige of academic and healthcare institutions and of scientific research, create a gradient of power that may not be favourable to the potential research recruit. An aim of research ethics, therefore, is to give power to the otherwise powerless – to make the subjects of medical research the masters of their own fate.

There are several core ways in which research ethics regulation protects individuals and gives them greater power. Any risks of research must be kept to a minimum and be proportionate to the anticipated benefits. Subjects of research must have given free and informed consent to take part; there must be no coercion, no undue inducements, and adequate information about what is required and what risks might be. There must be no negative consequences of refusing to take part (for example, no impact upon medical treatment). Subjects must be free to withdraw at any time that they consider the research too onerous, or indeed, for any reason or for none. Commonly too, there are requirements that the research must be scientifically valid and worthwhile. The Nuremberg Code was drawn up in 1947, and the World Medical Association Declaration of Helsinki was originally drawn up in 1963 and has been revised many times (Trials of War Criminals before the Nuremberg Military Tribunals under Control Council Law 1949; WMA 2008).

There are different levels at which research on human subjects is regulated. Some regulations have binding force, others are guidelines only such as professional guidelines or local practices. Some regulations deal with human subjects research generally; some deal specifically with certain aspects such as the regulation of the use of human tissue, DNA, or data pertaining to human subjects. There may be a degree of overlap and there is not always clarity between different sources of regulation. Many countries have their own laws regarding the conduct of research on human subjects; there are many legal instruments which together regulate research including laws about the handling of data, medical confidentiality, and privacy. Generally, national and local institutions have their own research ethics committees or RECs (Institutional Review Boards or IRBs in the US) which may include local variations both in the details of the regulatory codes and in the realities of practice. Ethical guidelines may sometimes go beyond what is strictly required by law.

2.2 An Equally Brief History of Ethical Issues in Genetics

It will enrich our inquiry to give some indication of why genetics, and latterly then genomics research, has attracted particular ethical scrutiny. The level of this scrutiny is again partly driven by historical events as well as by contemporary developments, and by views on the nature and power of genetics.

2.2.1 Worries About Eugenics

Historically, the idea that steps should be taken to improve the genetic "fitness" of the population has led to programmes of enforced or coerced sterilisation of certain people deemed "unfit to breed" – programmes of eugenics. Such ideas were promulgated by many who squeezed these conclusions out of doubtful interpretations of Darwin's account of the evolution, urging that we should give nature a helping hand in ensuring the "survival of the fittest". Prominent exponents included figures such as Julian Huxley, a leading member of the British Eugenics

Society, who did however argue against some more extreme proponents of eugenics. Fears that the least fit would breed excessively and destroy the health of the population in general were, ironically, very likely in many cases prompted by worries about sections of the population who were suffering not so much from deleterious genes as from the effects of poverty and the harsh living conditions of the industrial revolution. In certain places, these ideas were put into action: not just in Nazi Germany as is well known, with the forcible sterilisation of people with various disabilities, but also in some states of the United States and in other countries such as Sweden (Galton 2001; Kelves 1995). Many still express fears that modern practices in genetics may lead to such abuses or may amount to "eugenics by the back door".

Contemporary discussions of ethical issues in genetics often make specific reference to eugenics and to the practices of Nazi Germany – again often highlighting this one case – and in addition may specifically take steps to differentiate activities that might otherwise be seen to bear certain parallels to those of the past. For example:

> To avoid past abuses of genetics as exemplified by Nazi Germany, there is a desire in many countries to ensure that decisions to terminate an abnormal pregnancy are well-informed but not influenced by the view of others, particularly the state or other official bodies. The primary objective of genetic screening is usually seen as providing information to allow informed choices, particularly regarding reproduction ... Although carrier testing may result in many people opting to have pregnancies tested and affected ones terminated, this is not deemed to be an objective. Rather, this is seen as a consequence of the objective of the programme: informed decision-making (Marteau and Richards 1996).

Many have concerns that notions of eugenics may linger in genetics and genomics research. One worry that some express is that the findings of such research may be applied in ways that further a eugenics agenda.

Eugenics

The term "eugenics" was originally coined by Francis Galton, cousin of Charles Darwin, and has had many enthusiastic advocates as well as detractors (Galton 1892). Different supporters of eugenics have had varying ambitions, but all focus upon improving the general genetic "stock" of the human population, often drawing explicit parallels with the improvement of stock of domestic animals. The focus on genetic qualities distinguishes eugenics from other population-based health measures, which may, for example, aim to improve general health through diet, education, and sanitation. Strategies of eugenics range from encouraging certain sections of the population to reproduce, to discouraging others, forced sterilisation, forced abortions, or the eventual elimination of parts of the population, either as an envisaged result of various reproductive policies or directly through extermination. Policies such as attempts throughout much of the twentieth century in Australia to assimilate Aboriginal populations to the general population through forced adoption of Aboriginal children have also been seen as a form of eugenics, as may certain immigration programmes restricting entry to certain population groups.

(continued)

The United States first implemented eugenics policies in the early 1900s, prohibiting marriage and forcing sterilisation for those deemed mentally ill.

From the 1920s in Canada, sterilisation of the "mentally deficient" took place, with laws not being repealed until the 1970s.

In Germany in the 1930s and 1940s, eugenics programmes forcibly sterilised hundreds of thousands of individuals, and led to the direct killing of many. Positive steps to encourage the "racially pure" to reproduce were also taken.

Sweden practiced eugenics programmes and sterilisations on a scale second only to Nazi Germany. Following an investigation late in the twentieth century, the Swedish government started paying compensation to those coerced into being sterilised.

Belgium, Brazil, Norway, France, Finland, Denmark, Estonia, Iceland, Japan, and Switzerland also followed programmes of eugenics.

Eugenics fell into general disrepute following the revelations of the practices of the Nazis during the Second World War.

It would be an error to equate eugenics simply with policies stemming from state control. There is documentary evidence that within Germany official bodies of the medical profession were so enthusiastic about eugenics in the 1930s that following protests to Hitler in 1937, the activities of the medical profession in sterilising the "feeble minded" and others were reined in (Hanauske-Abel 1996).

Liberal Eugenics

"Liberal eugenics" is a term referring to practices of human selection such as pre-natal diagnosis using techniques such as chorionic villius sampling and amniocentesis, and pre-implantation genetic diagnosis, where these are presented as a matter of individual reproductive choice by potential parents rather than anything state-imposed.

Proponents of such a notion argue that the emphasis on individual free choice means that such practices are sharply divided from the state-sanctioned eugenics of the past, where individuals had little choice, or were greatly encouraged to participate, and where many human rights were violated. A generally liberal ideology of individual autonomy and freedom from state interference is used to support liberal eugenics.

The practices of genetic counselling and screening programmes for pregnant women tacitly support a notion of liberal eugenics insofar as claims are made that counselling should be nondirective and that the practices are designed to enhance autonomy and reproductive choice.

However far removed such contemporary practices may be from the forced sterilisations of the past, many critiques of such a notion of liberal eugenics have been made. For example, auditing of such schemes has often been in terms of termination of affected foetuses, which indicates an additional agenda from that of enhancing individual choice. It may be argued that there is an implicit officially sanctioned account of what characteristics are or are not desirable, enacted, for example, through professional and other standards of when a condition is sufficiently serious to justify the selection out of affected embryos or foetuses. The aim of reducing the number of individuals in the population affected by various conditions still seems to be in place; there are critiques of such aims, coming, for example, from some disability activists. Supporters of liberal eugenics may have no quarrel with such aims.

Perhaps a more worrying critique comes from arguments that there are constraints on allegedly "free" choice within healthcare settings, for instance, alleged bias in some material providing information on Down's syndrome. There have been various attempts to address such potential issues which are a problem for both supporters of liberal eugenics and those more sceptical of such practices.

A general issue for eugenics has been how the notion of an "improvement" to human stock should be understood; liberal eugenics may concentrate on the choice to eliminate certain more clearly demarcated genetic conditions, but discussion of "positive" choice in favour of certain characteristics such as sporting prowess and "good looks", or so-called "human enhancement" starts to raise issues closely parallel to those raised by the eugenics movements of old (Habermas 2003; Rothman 1988; Rapp 1999).

2.2.2 *Genetics and Populations: The Spectre of "Race"*

Closely linked to concerns about how eugenics causes ripples of worry about genetics are general worries about the denigration of sections of humanity falling short of elimination through genocide or constraints on reproduction through eugenics. In particular, worries about finding elements of racism within genetics are common (Nash 2011). For instance, the heated debates about claims that there are genetic links between different racial groups and IQ scores are well known. These worries about racism are also linked to notions of genetic essentialism – that race, an aspect of our identity, may be constructed out of genetic accounts of differences between groups of people.

However, despite the ease with which such worries may surface in relation to genetics, it is now commonly held that findings in genetics are actually disabusing the whole notion of race (Editorial 2001). Although broad differences between various population groups can be found, and although information about genetic ancestry is frequently important methodologically in genomics research, the idea that the human population can be divided into distinct racial groups is hard to sustain (Risch et al. 2002). Nonetheless, worries about race still regularly surface in relation to genetics. For example, some express concerns that certain research questions may be inappropriate. There is interest in the genetics of sporting abilities and some research suggests that those from different African populations may be more likely to carry genes equipping them for excellence in different sporting events (Blackburn-Starza 2011). Might such research reify concepts of race, or stereotype certain population groups? Some might worry about the use to which such research might be put; others may consider that such questions should not be investigated at all, regardless of what use is made of the findings.

2.2.3 Genetics and Reproduction

Many medical techniques in use and in development today involving genetics have produced great ethical concern. In many cases, ethical concern regarding research in genetics overlaps with concerns some people have about research involving human embryos and stem cells, concern felt especially by those who have certain views of the moral status of the human embryo and the foetus. Research into genetics has so far done a great deal to allow us to make finer predictions of who is at risk from various genetic conditions, but has done less to enable us to treat many of these conditions. Hence, a major application of genetic knowledge is through reproductive interventions. These include testing potential parents for carrier status, and importantly, techniques such as pre-natal diagnosis and selective termination of pregnancy, and pre-implantation genetic diagnosis. This link of the application of genetic knowledge with reproduction, and through practices which some worry are tantamount to eugenics, raises considerable concerns in some quarters, just as others hail these as welcomed breakthroughs. The involvement of reproduction also introduces some ethical issues of particular concern to women. These issues are all worthy of extended discussion, and are not the immediate focus of this book. Nonetheless, an appreciation of the strong links of genetic techniques with reproductive technologies is one reason for the intense ethical scrutiny that genetics and genomics receive.

2.2.4 Genetics, Life, and Identity

A major but often unarticulated reason why there is such concern about ethical issues in genetics is a powerful allure it seems to possess – genetics often seems special, particularly wonderful, particularly worrying, and precisely because of this ineffable allure, it is easy to make assumptions about it that cry out for careful analysis. (One of the principal aims of this book is to look carefully at research in genomics for what it is, and how it differs from or is similar to other research – to examine such allures for what they are.) One way of characterising the pull of genetics is that it seems centrally connected with "life itself". This connection is found not just in its obvious links with reproduction – what is passed from parent to child, from generation to generation, and with the techniques used for manipulating reproduction to control what DNA is passed on. It is also found in claims that genetics captures the essence of life, that it is at the heart of what it is to be human.

These claims are expressed in various commonly used metaphors that put genetics at the very centre of our being. The genome is said to be the "book of life"; the "code of life"; a "blueprint" for life; a "recipe" for life, a set of instructions that all cells in the body must follow. If this is correct, then it would imply that developments in our knowledge about genetics, and our ability to manipulate or act upon this knowledge, may well be of particular ethical importance.

Genetic Determinism, Reductionism, Genetic Essentialism, and Geneticisation

Genetic determinism is a doctrine about ***causation***. It holds that in order to understand the cause of a particular phenomenon, the best place to look is at genetics. This may be a general doctrine or specific to particular issues. It is possible to consider that some aspects of our nature are genetically determined and not others and to consider that some aspects are under a certain degree of genetic control that falls short of full determinism.

What has come to be known, following Francis Galton, as the Nature/Nuture controversy, pits genetic cause against environmental (or social) causes. If you consider that, for example, intelligence is determined genetically, then that would exclude social and other environmental factors that others consider have a role to play in determining intelligence. In reality, for many such traits, it seems a certain proportion of the causation is due to genes, the rest due to other factors.

Genetic determinism is a narrower view than biological determinism. Genetic determinism considers that genes are the primary biological cause, over and above other biological causes.

A calculation of the degree due to genetic causes is known as the "heritability" of a characteristic, expressed as a percentage where a score of 100% would indicate that a trait was entirely governed by genetic factors.

As our understanding of genetics grows, scientists however are more and more observing that there are other influences, and that the interaction between an individual's genome, their environment from conception throughout life, and even intergenerational factors all interact in a highly complex manner. It is probably fair to say that popular accounts of genetics more readily adopt a degree of genetic determinism than more nuanced scientific accounts.

Genetic reductionism is a view that the best ***explanation*** for certain phenomena, such as health or behaviour, can be given by reference to our genes. It too can be seen as a narrower version of biological reductionism.

(continued)

A genetic reductionist account might be given for example of mental illness or other behaviour if this was thought the best or the primary mode in which this could be explained. Ambitious programmes for genetic reductionism include attempts to explain criminality and religious belief with reference to an individual's genetic makeup.

Genetic essentialism is a view about our ***nature*** or essence, about who we are. It holds that genetics holds the core of our identity, and in understanding genetics, we capture the "essence" of who we are.

It can be contrasted with other possible accounts of our nature. It could be held that there is no one thing which encapsulates who we are, but that our makeup is formed from a myriad influences including biology, our social identity, and our culture and history. (The idea of "identity" is a highly complex notion much discussed by social scientists and others.)

Other alternative accounts include the idea that each of us has a soul which captures our essence; that we are formed in the image of God; and that we have no essence – as existential philosophers such as Jean-Paul Sartre have argued (Sartre 1973).

A phenomenon which can be seen as the spread of a genetic reductionist or essentialist approach was described first in the early 1990s by Abby Lippman:

> **Geneticisation** refers to an ongoing process by which differences between individuals are reduced to their DNA codes, with most disorders, behaviours and physiological variations defined, at least in part, as genetic in origin. It refers as well to the process by which interventions employing genetic technologies are adopted to manage problems of health. Through this process, human biology is incorrectly equated with human genetics, implying that the latter acts alone to make us each the organism he or she is (Lippman 1991).

There have been controversies about how geneticisation should be understood, how problematic it might be, and how widespread any such phenomenon might be (Moore 2001; Marks 2003; Wilkie 2001).

2.3 Meet ELSI

Streams of research examining ethical, legal, and social implications of scientific research, particularly in genetic and genomics, are now common. The particular form of these ELSI programmes stems from the doings of James Watson whose first act as director of the Human Genome Project at the NIH (National Institutes of Health) in the USA was to announce, apparently without discussion, that 3–5% of the HGP budget was for ELSI work. Especially when you consider how expensive such research is, and how relatively cheap ELSI work is, requiring little in the way of equipment compared to genomics, this really amounts to a large sum of money. A background to this announcement had been a long-standing conflict between scientists and those in the US wishing for a moratorium on recombinant DNA research. James Watson himself was concerned about the implications of genetics research, and many likewise were worried about the implications for society of the newly unfolding findings in genetics. Other major research funding organisations such as the Wellcome Trust and the UK Economic and Social Research Council also set money aside into research into ethical legal and social aspects of genetics and genomics.

The history of eugenics abuses is thus one major reason amongst others for the current emphasis on examining ethical issues in genetics. Paradoxically, it may also be one reason for opposition to such ethical scrutiny, since who thinks of themselves as doing anything as bad as that? It is a fair point that an idealistic research scientist, motivated by disinterested love of science or by the wish to contribute towards beneficial outcomes for humanity, may feel slighted to be thought of in the same category as those who tried to eradicate sections of the population or to prevent the "substandard" from reproducing. Likewise, worries sometimes expressed in the popular media that by "fiddling with humanity's essence" scientists are "playing God" is likely to be met with a sceptical sigh by a scientist facing yet another routine day in the lab examining a minute portion of the three billion base pairs that make up the human genome.

Warning to the Complacent: A Touch of Irony – The Values of a Failed Organisation

ELSI work and the regulation of research ethics in general draw upon an obvious attraction to the idea of a dedicated programme to look at ethical issues within an enterprise or organisation. But organisational attention to ethics is not a failsafe strategy. The futility of certain value statements is rarely more clearly illustrated in this fine sounding visionary list:

"OUR VALUES

RESPECT: We treat others as we would like to be treated ourselves. We do not tolerate abusive or disrespectful treatment. Ruthlessness, callousness, and arrogance don't belong here.

INTEGRITY: We work with customers and prospects openly, honestly, and sincerely. When we say we will do something, we will do it; when we say we cannot or will not do something, then we don't do it.

COMMUNICATION: We have an obligation to communicate. Here, we take the time to talk with one another . . . and to listen. We believe that information is meant to move and that information moves people.

EXCELLENCE: We are satisfied with nothing less than the very best in everything we do. We will continue to raise the bar for everyone. The great fun here will be for everyone to discover just how good we can really be."

This is taken from Enron's 1998 Annual Report. Not only did the company fail, when it went, it took with it the livelihoods, pensions, savings, and investments of thousands and thousands of ordinary hard-working people, as well taking with it the previously well-respected firm of Arthur Anderson, and producing a vast knock-on effect on the financial system. Those who

(continued)

examined what happened at Enron argued that mismanagement, greed, neglect, and corruption were so rife within Enron as to be the norm (McLean and Elkind 2004).

2.4 Too Much Ethics?

Is genetics getting too much scrutiny? Is this degree of emphasis on ethical issues in genetics really justified? In response to such claims that genetics is special, and thus potentially raises acute ethical problems, others argue equally vociferously that genetics is no different to many other areas, and that undue attention to it can even be harmful. Is there a danger that we are falling into "genetic exceptionalism" here in singling out genetics and genomics?

Germany's history has led to the country having distinct legislation on matters relating to genetics and research which are particularly restrictive. Some argue that this is an example of overreaction to past problems. James Watson again: "For almost two decades the totally unjustified German public perception that recombitant DNA manipulation *per se* is inherently evil has led to much unnecessary if not ludicrous regulation and legislation that has greatly set back the country's biological and medical research, as well as long stifling the use of DNA technologies by German industry, be it large or small .... German opponents of DNA have misleadingly raised the moral threat of Nazi-type geneticists working to overturn the natural order of life. In letting hysteria triumph over the use of reason, recombitant DNA products like insulin cannot be industrially produced in most parts of Germany" (Muller-Hill 1998).

As part of my work with the Procardis partners, in April 2009 I talked to a group of the researchers based at the Maria Negri Institute in Milan about their views on a range of questions in relation to ethics and genetics. This was an interesting group who appeared to have thought about these issues and come to considerable consensus. The whole group rejected very strongly the suggestion that there were any particular ethical issues arising with genomics research.

My approach to the issue of genetic exceptionalism is that it is indeed sometimes the case that claims are made for the uniqueness of genetics that do not bear close examination. For instance, it is sometimes claimed that genetic information is unique in that genetic information about one person has implications for other family members. However, so too does TB infection, scabies, and many other infectious or parasitic diseases, which raise the risk for those in the same household; information about the immediate environment, and about the social history of family members, has implications for others just as genetic information does. What we need to do is to see clearly where there are differences and where there are similarities.

One problem with overstating the uniqueness of genetics and then linking this to an elevated level of ethical difficulty is that this may tend to alienate those working in genetics who feel they are unfairly subject to ethical scrutiny that others escape; and as I have argued, close dialogue with those on the ground is vital to an effective appreciation of ethical issues. Scientists with knowledge of genetics may well feel that many claims in popular literature for the exceptional status of genetics are exaggerated; for instance, claims that DNA is an exact determinant of how an individual will develop are based on highly simplistic biology that is demonstrably false.

Another problem with important methodological implications with overstating the "uniqueness" of genetics is that useful comparisons and insights from other areas may be overlooked. Yet using such comparisons can be an indispensible tool in enriching one's thinking. For example, identifiability of data is a general problem in many areas of research. It is often said to present particularly pressing or even unique problems in genetics because extremely large quantities of data about individuals can be generated and shared very readily. However, surprisingly similar problems can arise with other research. Much work in the social science can produce data about individuals, families, and groups which can present problems when trying to comply with ethical standards protecting privacy and anonymity. For example, there are postal areas in Australia where only a few Aboriginal families live. Data on ethnicity and postcode could in theory easily identify an individual family.

What is noticeable about some discussions is that it can quickly become a bragging match; problems in genomics are worse, more pressing, more urgent, and more insuperable. This is of no use to anyone. We need to look carefully at what is in common and at what is unique. There may be for example technological solutions to the protection of data that could be shared between different research methodologies and those that may not so easily be so shared.

It is also sometimes arguably the case that some ethical issues arise because developments in technology and in scientific understanding can greatly increase power and capacity to achieve certain things (see Sect. 5.4). The area of genetics and genomics is replete with such powerful and rapidly developing technology. But the only approach which stands muster is to see genomics for what it really is, compared and contrasted as accurately as possible with other fields of endeavour, the more accurately to gauge how best to respond. (This in fact is part and parcel of developing the *phronesis* discussed in Chap. 1.)

Thinking of genetics as "unique" or as providing particular ethical problems is in any case of scant practical use, given that we need to understand precisely what problems we need to tackle. Our guides to action must derive in some way from other areas, and we need also to strive for some notion of consistency – as shall be discussed at greater length presently. We will consider such matters when we consider how problem areas differ one from another – how consideration of genomics research might bring in new issues compared to other research. We also have to consider similarities and differences within genetics and genomics very carefully. This will be a theme that will recur throughout the book.

2.5 Human Subjects and Subjects of Research

> Act in such a way that you always treat humanity, whether in your own person or in the person of any other, never simply as a means, but always at the same time as an end.
> "Formula of Humanity" (Kant 1948).

How and why do ethical issues arise in medical research? In considering ethical questions, one needs to stretch one's imagination, to walk slowly around the object of inquiry, as it were, looking at it from different perspectives. Here is one such walkabout. There is an interesting ambiguity of the English language that the word "subject" has multiple meanings. In one of these, it can be used to refer to human subjectivity, to the way in which we humans are subjects of experience, in contrast with mere objects. The Oxford English Dictionary gives this as its ninth meaning for the noun "subject":

9. *Mod. Philos.* More fully *conscious* or *thinking subject* : The mind, as the "subject" in which ideas inhere; that to which all mental representations or operations are attributed; the thinking or cognizing agent; the self or ego.

As such, it refers to an especially precious feature we share, for only thinking, sentient beings can be subjects in this sense. In this sense, it may be contrasted with "object", meaning a merely physical thing. As such, it can act as a grounding basis for explanation of ways in which we have value and merit ethical treatment. But "subject" can also be used in very different ways. It could mean the subject of a realm, one bound by obligations to obey another. Besides having one sense in which it is *contrasted with* objects, in another sense, it can *refer to* objects, to physical things which are operated on in some way, as in the practice of a craft. Relevantly for our purposes, it could also mean "the subject-matter of an art or science", and this can include:

e. A body used for anatomical examination or demonstration; a dead body intended for or undergoing dissection.
f. A person who presents himself for or undergoes medical or surgical treatment; hence, one who is affected with some disease.
g. *Psychical Research* A person upon whom an experiment is made.

This digression into semantics is used to make us ponder the intersection of different ways of being a subject, and a central reason for moral tensions in the conduct of medical research, stemming from different ways in which a "subject" of research may be simply a subject – meaning in one sense an "object" – and how to ensure that "subjects" of research remain "subjects" in the full human sense.

In fact, the word "subject" in relation to those who take part in research is now used much less frequently precisely because it carries the connotation of being passive, of being acted upon. The word "recruit" or "participant" is used preferentially by many; to participate in something implies a more active role, and therefore may be more respectful. But of course, those who take part in research do still have things done to them, and it is important that the language we chose reflects reality.

One might with some justification consider that in much research the active role suggested by "participant" is a trifle optimistic.

This point about language is a means of thinking about one avenue through which medical research sometimes attracts ethical scrutiny: on the one hand, medical research is an obvious good, since it is aimed at improving medical treatment and prevention strategies (and undoubtedly, this is one reason why many feel a certain impatience with the degree of attention paid to issues of ethics and regulation in research). However, doing medical research on humans brings into focus a key feature of our existence as embodied, material beings, as exemplified in the ambiguities of the word "subject": so we ourselves can therefore also be studied "perhaps almost as narrowly as a man with a microscope might scrutinise the transient creatures that swarm and multiply in a drop of water" (Wells 2006).

There is great resistance to treating someone "as an object". But in an important way this is precisely what medical research has to do. Of course, the practice of clinical medicine also of necessity acts upon an individual's body; there have also been attention to ways in which medical practices may in doing so dehumanise their patients, but the proper practice of medicine is inherently aimed at caring for the individual patient. Medical research, on the other hand, sometimes has the possibility of offering benefit to the subjects of research, but more often is geared towards helping others (and much research is a long way from clinical application). Resistance to objectifying humans is a large part of what lies behind the application of the notion of "respect for persons" in medical practice in general and in research in particular: for a fuller discussion of respect for persons, see Chap. 8.

Kant's "formula of humanity" with which this section started has historically been an extremely important source of inspiration for various expressions of the requirement to respect persons. It is also, interestingly for us, onc of the most misquoted parts of philosophy: generations of philosophy students err in claiming that Kant said we should never treat people as a means to an end. Think of this for a second, and it is the end of all human society. Kant more precisely claimed that we should not treat others *merely* as an end to our own purposes. *Working out how we can use others in research without treating them merely as an end is perhaps the central task of research ethics.*

This difficult task is compounded perhaps by a sometimes perceptible current of resistance (indeed occasionally revulsion) towards the notion of treating people as the subject – the object – of scientific study. Whilst this may seem utterly irrational and misplaced by those with an interest in conducting and benefiting from such research, those who are heirs to the Enlightenment tradition of scientific inquiry as a key way forward for the human race, I believe it is important to be aware of the strength of such a response, a response shades of which may be found within thinkers from the Romantic tradition which followed as a reaction to the rationalism of the Enlightenment.

An illustration of an extreme attitude to the treatment of a human being simply as an object of research is examined in Georg Buchner's play *Woyzeck* (Buchner 2004). Woyzeck is a soldier who is portrayed as of low social standing, treated with disdain by those "above" him. He is used as the subject of medical experiments, including one where he is put on a diet consisting only of peas. It becomes apparent throughout the play that his mental state is deteriorating badly. The setting of this medical experiment, in a context where the subject is viewed as a minor being of lesser moral standing than his social and military "superiors", presents very starkly both a vision of disdainful treatment of research subjects and perhaps a view of the whole enterprise of researching on humans as in itself part of an ethically dubious stance towards one's fellows.

One clear example of this could be seen in some of the reactions in the UK to what became known as the "Alder Hey scandal". The following discussion draws on work analysing the report on this together with Steve Dewar of the King's Fund (Dewar and Boddington 2004). Towards the end of the last century, there was an unexpectedly high number of deaths of children undergoing heart surgery at the Bristol Royal Infirmary. During the course of an ensuing inquiry into these deaths, it was mentioned in evidence that there was a collection of hearts at the Royal Liverpool Children's Hospital. It probably surprised many involved in medical practice and research that the existence of this (actually world-renowned) collection was news. However, an inquiry into the collection was started. It was found in due course that this collection of hearts was in fact well known, well managed, well used, and instrumental in saving hundreds if not thousands of lives. However, in the course of the investigation, another collection of children's organs was discovered at the Royal Liverpool Children's Hospital at Alder Hey, a collection found to have various ethical shortcomings.

The point I wish to make about the ensuing furore and the inquiry that then took place into the organ collections at Alder Hey is not about the details of right or wrong practice regarding this other collection. There are multiple facets to this case. There was evidence of mismanagement, and of lack of resources meaning that collections were not curated or utilised to a very high standard, and there were problems with levels of information and consent with some of the families. However, there was also (in my opinion) an element of scapegoating of some concerned. These issues are important but they are not my immediate concern here. A fascinating, and disturbing aspect of both the inquiry and surrounding reporting and reactions was what could be seen as a streak of anti-science, anti-research running through responses to the case.

Much that has been written about this, especially in the media and in the initial "revelations" about the collection, suggested an element of surprise that such organ collections are necessary. But how else is research to be conducted? Much that was written also carried the implicit suggestion that researching on human body parts was inherently wrong, and that the onus of proof was on medical research to show that it was not being abusive or immoral. There is of course a need to obtain such material properly. But given that medical research aims at improving health, one might reasonably quarrel with the suggestion that examining the dead human body is in itself problematic. (This discussion puts to one side those religious beliefs that prohibit such examinations.)

In the inquiry and its aftermath, there were strong links of medical research, especially the use of body parts, with the macabre. The inquiry report specifically includes photographs of the two collections it investigated (Royal Liverpool Children's Inquiry 2001). One was the heart collection which sparked the inquiry, a collection of international standing, frequently utilised by those performing life-saving surgery, and deemed satisfactory. It happened to be housed in a modern, white, clean-looking building. The problematic collection just happened to be housed in the basement of a gothic-looking ramshackle Victorian building in Myrtle Street. The inquiry report stated that it included these photos to "give some idea of the nature of the building" and that the professor involved "enjoyed the seclusion that Myrtle Street afforded him"; reading the report, it is hard to see what purpose the photos hold, other than in producing a classic "Frankenstein" picture of gothic horror.

There were particular psychological challenges to the subject matter of the Alder Hey Report: it dealt both with the dead and, not only that, with dead children: the loss of a child is particularly hard to bear. Respect for the dead takes various forms but seems to be universal in human cultures and is one of the hallmarks of the very earliest human civilisations. But Alder Hey is an instructive lesson for what was revealed about underlying attitudes to the use of human tissue and organs as objects of scientific research. Some elements of similar resistance to research in genetics – to "cracking the code of life" – can also be discerned in various places. Such sentiments of the clash of science with humanity are also found, for example, in Charles Dickens' attacks on utilitarian approaches to counting and measurement in relation to human beings, for example, in the character the headmaster Thomas Gradgrind in *Hard Times*: "A man of realities. A man of facts and calculations. . . . With a rule and a pair of scales, and the multiplication table always in his pocket, sir, ready to weigh and measure any parcel of human nature, and tell you exactly what it comes to. It is a mere question of figures, a case of simple arithmetic" (Dickens 2009).

The dangers of an unfettered "objectivity" are expressed again here in relation to Nazi practices: "Scientists espouse objectivity and spurn value judgements. But pure objectivity leads to regarding everything as feasible. The killing of mental patients? If it is objectively necessary on economic grounds, why not? The use of mental patients, the Jews, and Gypsies, as experimental animals before and after their necessary deaths? If the authorities allow it, why not?" (Muller-Hill 1998).

Utilitarianism

Utilitarianism is a highly influential moral theory that claims that the rightness or wrongness of an action consists entirely in an assessment of its consequences – together with an account of what consequences we should aim at. In its classic form, it aims to produce "the greatest happiness for the greatest number", and aims to maximise happiness or pleasure, and minimise unhappiness or pain, over the population as a whole. Utilitarianism implies that the ends at which we aim can be measured and compared. Many different forms of utilitarianism have developed, including forms that claim we should try to maximise not happiness, but the satisfaction of desires.

(continued)

Utilitarianism has its roots in philosophers from the end of the eighteenth century and throughout the nineteenth century, importantly Jeremy Bentham and John Stuart Mill (Bentham 1823; Mill 1999). Utilitarian social reformers of the nineteenth century undoubtedly helped to change our social and moral climate, arguing, amongst other things, for more equal treatment of women and more humane treatment of workers, of children and of the poor. Mill also argued forcefully in favour of free speech. Utilitarian ideals are still firmly embedded in social policy insofar as it looks at the outcomes for all those affected, although other elements are often present too in social policy that may place limits on a consequentialist approach. Utilitarian thinkers are also still influential in contemporary debates about ethics. Perhaps the most prominent of these currently is Peter Singer.

Many other accounts of ethics include elements of attention to consequences. For example, the principles of beneficence (aiming to give benefits) and of nonmaleficence (avoidance of harm) are consequentialist in form, and many hold these principles together with others such as autonomy and justice.

Critics of utilitarianism often argue that there are some things we should never do, no matter what the consequences are. For example, suppose a research project could proceed faster and more cheaply without asking patients if their samples could be used in the research, and cause little harm: this would be held by many to violate respect for persons, and hence wrong, even if the overall outcomes might be best for humanity as a whole.

The retention of a humane attitude alongside the dispassionate examination of the human subject is essential. *What is crucially needed is a detailed account of how we can retain an attitude of humanity to each other*, especially in contexts where the person's body is the focus of scientific concern. The furore surrounding the Alder Hey organ collection demonstrates this: clearly one thing that did go wrong was that there were competing notions of what it was to treat the dead with respect, together with problems of communication and understanding. We will discuss in greater detail later in this book how humanising and dehumanising factors may operate for good or ill (see Chap. 8).

2.6 Lessons from Research Abuses and Open Debate

We have looked at the historical background to research ethics and the scrutiny of ethical issues in genetics, much of it spurred by the abuses of the past. These abuses stemmed from a world view that saw some humans as inferior to others, sometimes as so inferior as to be expendable; a world view that saw individuals not as subjects but objectified them, reducing humans to bundles of hard facts including on some occasions alleged facts about their racial or genetic origins. Such abuse reduced human subjects to objects. We have seen how the responses to such abuses take the form of regulations and oversight designed to ensure that the participants in research are treated with respect and human dignity, and do not suffer any avoidable harms. Regulations act to ensure that individuals and groups are treated fairly, and that no one is singled out for substandard treatment. This is quite right.

But there is another equally important lesson to learn from past abuses, especially those of the Second World War: that what was wrong was, crucially, an intellectual regime that allowed such ideas to gain ascendency: a regime which relied upon the curbing of debate and of the free exchange of ideas and opinion. Albert Speer, Hitler's architect and Minister for Armaments, has written of "the stifling intellectual atmosphere of the Third Reich" (Speer 1970). One rationale for giving the research subject the right to *free* and *informed* consent is to counter such control of information. But the lesson we need to learn is much wider than this and requires attention to the broader context in which beliefs, choices, and actions are formed and made. This is a crucial point; indeed, it is one of the reasons behind the writing of this book, to encourage and enable debate about issues of importance. It is also a point to which we will return, for example, in discussing the importance of working within institutions which allow and enable dissent and open discussion of ideas.

Is there a guarantee that in a democratic system such problems could not arise? No. Programmes of forced or coerced eugenics also occurred within democratic political systems. More is needed than just the bare bones of democracy. Perhaps one essential ingredient is a widening participation in debate. Those who were the subjects of eugenics programmes, or their advocates, were not a party to the decisions to instigate these programmes. We should therefore see the imperative to democratic involvement and free and widening debate and discussion on these issues. This applies to all aspects of life, and research ethics and genetics should be no exception. In fostering such debate, it is important that it be as high a standard as possible. This book hopes to be a small contribution to fostering such open and high-quality reasoning. For further discussion on how problems might nonetheless arise even in the presence of ethics regulation, the reader is directed to Chap. 7.

2.7 Researchers Facing Ethical Issues: Accounts from the Front Line

This chapter has been discussing the need to examine ethical issues in genetics, in research, and in genomics research. This can be explained on both theoretical and historical levels. But there are also plenty of personal accounts of situations where researchers are faced with ethical issues that demand more than simply "getting ethics approval" or following a set protocol. This is an inevitable feature of real life which is always full of complexities. A set of rules and regulations may not provide an entirely unambiguous guide to action: the regulations are not railway lines that tell us exactly where to go and what to do. We may find that two rules clash when we did not expect this; we may find that we can now do things we had not regulated for; we may find that we do not know how to classify new situations and events. In an area like genomics research, the very pace of change also means that unforeseen situations may develop that challenge the current ethics guidelines, and at such times the different values involved leave plenty of scope for genuine disagreement over what to do. Crucially, in many instances, moral questions are impossible to avoid, since to fail to act is in itself a moral response.

Much research regulation has responded to examples of poor behaviour. But there are many situations where researchers have acted in ways that go beyond what is strictly required of them. To use a technical term from philosophy and theology, to go over and above the call of duty is to perform an act of supererogation. Such cases also amply show how ethics in research is not to be simply equated with the requirements of ethics regulation (see Chap. 3). For example, a former colleague had ethics approval for a project that involved contacting individuals with a certain diagnosis of a genetic condition from records of a hospital department. In all cases, the diagnosis had been made several years previously. It quickly became apparent that the individuals she contacted had either never been told of the diagnosis or had forgotten it. She considered that it was not her place as a researcher to inform these individuals of the diagnosis through her research project, and abandoned that part of the study, making her own judgement that to continue would be unethical. Let's now look at some examples of researchers in genomics facing ethical decisions.

2.7.1 Researchers Behaving Well: Detective Work Amongst the Mormons

Jane Gitschier conducted research designed to test the limits of security of stored genomics data (Gitschier 2009). The essence was to test how data could be triangulated: in other words, how data from different sources could be matched up to find out extra information. Gitschier combined data from the extensive Mormon records of ancestry with genomics data from the HapMap project which included many people from an area with a large Mormon population, and by using information gained from the Y chromosome (which is passed down almost unchanged from father to son) found herself in a position where it would have been possible to make educated guesses about the surnames of some of those present in the anonymised HapMap database. At a certain point in her project, she was faced with the question of whether or not her use of genomics data violated the terms of informed consent for participants in the HapMap. I emailed her a number of questions.

PB: I wonder if you could provide me with some clarification of a point in your paper. You state "in consultation with investigators at the University of Utah, where the samples were collected, we jointly concluded that confirmation of the predicted surnames would violate the ethical constraints of informed consent obtained during the collection of these samples because the names of the subjects would be used in the analysis."

Jane Gitschier: This was very interesting. When I did the work with the Sorenson database and realised that I was going to have a good chance of "naming names" I first contacted Ray White (because of course he was at Utah at the time of the collection) and he referred me to Mark Leppert who is still there. So I contacted Mark, and with much persistence we finally had several conversations, including with their resident ethicist Jeff Botkin. I asked them whether it would be possible to confirm the predicted names (I believe I sent them table 2 with the predicted names in a column). I wasn't even necessarily interested in saying precisely which ones were correct, but rather, say, what percent of them were correctly identified, which struck me as a reasonable compromise. However, after a near eternity (OK, about a year), I finally was able to get a conclusion from Mark, who said that even looking into the records to associate a HapMap ID number with a name was violating the consent agreement.

PB: It is quite clear that you were acting in the best interests of the subjects: is it possible to tell me if you concluded that this was a clear implication of the ethical constraint of informed consent, or how far you were making a cautious interpretation to protect the subjects?

JG: I think it became clear to me in talking to Mark that this was an ethical constraint of informed consent. Frustrating a bit from the point of view of validating what I was trying to show, but ethically it was absolutely the right way to go.

PB: Is it possible to supply me with the relevant informed consent material?

JG: All of the Utah subjects were re-consented for their inclusion in the HapMap project. I've attached the consent form. You'll see under risks that they do mention that identity might be compromised by comparison with data from other databases. So it was good that they had that in there.

PB: You also state "Moreover, in deference to the privacy of those who contributed the CEU samples [samples from Utah residents with Northern and Western European ancestry] I have not included the predicted surnames in Table 2." Again, I was wondering if you thought of this as doing more than was strictly required to protect subjects' interests. In other words perhaps whether you consider that a less scrupulous researcher might have been able to claim that the presentation of surnames was not actually in violation of strict requirements.

JG: Regarding this, I had several long discussions with my colleague Dennis Drayna who was Ray White's postdoc when these samples were originally collected 25-plus years ago. Dennis felt very strongly that I should not include the names (his exact words were "You can't do that!") As tantalising as it was to do so, I ultimately felt that he was right. I don't believe that Mark Leppert had an opinion on it or whether I discussed it with him. My memory is foggy on this, but I think Botkin concurred with Dennis.

PB: Your paper attracted my attention for many reasons not least because it seemed to me an example of a researcher going beyond a strict interpretation of their duties.

JG: When I was about to publish this, one of my colleagues here said, "Now let me get this straight – you're like a computer hacker who is hacking into a programme just to show the ways it can be hacked." And yes. That is what I did. It was really an amazing process!

The other thing that perhaps I should emphasise is that the initial point of the study was the simple question of whether any of the HapMap individuals could trace their ancestry to one of the Mormon founding fathers. As polygamy is one of the signatures of the early founders, and as the surnames of Young and Smith are very common, I thought it might be possible ask that question. This question came up in my mind because I was working in Chris Tyler-Smith's lab on sabbatical; he and his colleagues had demonstrated that they could make inferences about Genghis Khan progeny, for example.

I discovered that not only could I ask and answer that question, but I could make what I thought were reasonable inferences about the surnames of most of the male participants – the names simply popped out during the analysis. I recall this was a rather scary moment, and I immediately realised I was on the cusp of an ethical dilemma. Should I press forward? As there was so much discussion about genetic privacy in journals and meetings, I decided to do the analysis to prove the point that even a person like me, without super computer skills, could, with a lot of patience and care, figure this out. Also, to be blunt, I was getting tired of all the Poo-bahs in my field speaking *ex cathedra* about how their security protections were going to work. There is a vast world of genetic genealogy out there that does not fall under research restrictions, and nobody was thinking about it!

It is interesting to note the long process of getting advice and opinion on what was permissible. The HapMap consent forms are in fact very detailed, and in fact do specifically state that: "We will protect your privacy in several ways. While [insert name of institution] will keep your signed consent form, nobody else will see it. We will not keep your name with your sample or give your sample a code number that could identify you. So nobody at the Repository or who studies your sample will know that it came from you" (The International HapMap Consortium). In this case then, it seems reasonable to interpret this as meaning that matching names to samples would violate the terms of participation in the project. However, consent forms for different genomics research projects can be variable, both in level of detailed information and in the content of what is agreed by both parties, so in other instances, such a matching may have been consistent with the precise terms of the original consent. Some national laws on data protection may allow a wider use of data if this is construed as "fair". This therefore is an example of how challenges may arise which might not always be covered by the exact wording on consent forms.

How might one have held an alternative view? The very remit of what Jane Gitschier was trying to do might have led to a different interpretation of what was permissible. The remark of her colleague that this was like being a hacker helps to orient us to a different view (and is an illustration of the general strategy that mulling things over and hearing the points of view of others can help us to enrich our moral thinking). Her aim was not actually to *hack* – if this is understood as aiming to use any information illicitly, in the sense of using it against the interests of participants – but to demonstrate its possibility in principle. Such a demonstration would be of great use and interest in designing safeguards for participants in genomics research, and indeed for customers of genetic testing companies including ancestry testing, who may have a great interest in having their privacy protected. It might be reasoned that at least going as far as ascertaining how often the guessed surnames were correct would lend scientific credibility and validity to Gitschier's work, and that this in itself was a value to weigh up against other values. If one considered that the *spirit* of the consent agreement that the participants signed was aimed at the protection of their privacy from illegitimate interests, or from malicious hackers, one might argue that allowing this "in principle" checking of the accuracy of surname guesses would be consistent with this spirit, since it could assist with assuring privacy of participants in the face of increasingly powerful technology and statistical techniques. One might possibly argue that the wording of the consent form implies that when it was claimed that no-one studying an individual's sample will know whose it is, this was simply an implication of the undertaking not to keep names with samples. Hence, perhaps a project that precisely aimed at examining how well anonymisation procedures work to protect privacy was in keeping with the general spirit of the consent.

The International HapMap Project

The International HapMap Project started in 2002 involving several different countries in an effort to identify genetic variation in humans. Genetic sequences from individuals from different population groups are compared to identify chromosomal regions of shared genetic variants. All of the information generated by the Project is available in the public domain. Releasing this information to researchers enables them to find genes associated with disease or pharmacological differences between people.

The HapMap project also involved community consultations to explore potential ethical issues in population genomics research.

2.7.2 Researchers Behaving Well: Warning the NIH About Privacy Risks

In the summer of 2008, a group of researchers found a crack in the edifice of privacy protection for genomics research subjects (Homer et al. 2008). Using highly powered statistical techniques, these researchers managed to show that, given possession of an individual's DNA, it was possible to demonstrate if that individual, or a close relative, was present in a sample of 1,000 individuals whose DNA had been aggregated. The core idea behind this technique is simple enough. For any given population, any trait that can be measured has an average, or a particular rate of occurrence. Each individual stands in a particular relation to the average for each of these traits. For genomics which measures variations known as single nucleotide polymorphisms (SNPs) in individuals, each SNP has a certain rate of occurrence in the population, and each individual has or does not have a particular variant. Given an aggregated group selected from within this population, if there are enough traits measured, then, for each of these traits, it is possible to calculate the probability that an individual is present in that group by comparing the average of that group with the individual's DNA for each of the multiple traits in question. This possibility had been known about; it is just that most people didn't think you could actually do it, until Homer et al. showed that this was in fact achievable.

Did they rush to publish as dedicated researchers generally will with new and exciting findings? No. Realising the potential problems for the promises of privacy and anonymity given to research subjects, they did the responsible thing and firstly contacted the US National Institute of Health (NIH) to warn them of their findings before publication. It is not clear that there are any standard ethics committee guidelines which would have mandated this warning. It sprung rather from independent and responsible action.

2.7.3 *Researchers Behaving Well: What Watson Didn't Want to Know*

James Watson was one of the first people to have his complete genome sequenced (Olson 2008). (In the initial sequencing of the human genome, a public consortium sequenced a mixture of several people's DNA, a private consortium sequenced one man's DNA – the owner of the private company Celera, Craig Venter.) Watson's DNA sequence has been published in full, except for one small part of it, the stretch of sequence for the Apo E gene, which would give some indication of Watson's risk of developing Alzheimer's disease. His grandmother suffered from the disease: Watson himself said he did not want to know what his risk was.

But such is the power of DNA that much can be gleaned from even incomplete information. There are statistical relations between particular variants within the genome, such that it is possible to infer from some variants what other variants an individual possesses. Researchers based at Brisbane in Australia demonstrated that from other known patterns of variation within Watson's DNA, it would be possible to work out his ApoE status and therefore give an estimate of his risk of Alzheimer's disease (Nyholt et al. 2009). This could even be done, they demonstrated, using publicly available data. They published the broad findings of their technique, but did not estimate of Watson's ApoE variant. They proved the possibility of this through examination of J Craig Venter's DNA, which is available publically in full. Indeed, they contacted Watson to warn him of their hypothesis, again like Homer et al. before the publication of their paper, and following this, a larger section of Watson's DNA sequence was withdrawn from public access.

2.8 Epilogue: What Can We Learn from Good Behaviour?

Standard accounts of the history of research ethics regulation orbit around cases of abuse and malpractice. But for a full account of ethics, looking only at what goes wrong is a pretty lopsided approach. From these examples we can see the virtues of researchers taking steps to warn funding bodies and regulators (and likewise, of such bodies listening); of responsible publishing etiquette; of the application of general rules of privacy, what might be called "common human decency" and of careful interpretation of ethical rules, of discussion with colleagues. All thought independently and some acted beyond what they could have strictly speaking "got away" with. This then links in very strongly to the above discussion of the need for free and open debate and for independent thinking. There are also reasons to suspect that too much attention to wrongdoing alone is likely to lead to increases in rules and regulations in response and there are some reasons to worry about this (see Sect. 3.2.4). We will ponder such issues at various points throughout the book; meanwhile, a general lesson to take with us is this: draw up guides to ethical behaviour not just from bad behaviour, but also from good.

2.9 Conclusions

There are various reasons why there has been a great deal of attention to ethical issues in genetics: what is important is to look closely at what is the case and see where we might gain insight from other areas or where particular difficulties might lie.

It is morally important to give voice and power to the subjects of research.

However useful they might be for instilling standards, codes of ethics on their own cannot be enough: abuses can still occur even where regulations exist.

A crucial failing that led up to some past abuses in research ethics was the lack of democratic free and open debate; there is a need for this within research institutions, and within the public arena in general.

For ethical guidance, we need to look not just at bad behaviour, but at good behaviour as well.

2.10 Study Questions

1. Would it be okay to work out James Watson's risk of Alzheimer's disease, as long as you kept this information to yourself? Would it be okay to work this out after he was dead?
2. What does it mean to treat someone merely as a means to an end? Try to think of a few examples and try to identify what exactly is ethically problematic. Think of examples which differ from each other. For each example, try to work out what would have to change to make the treatment ethically acceptable. See also Chap. 8.
3. From your own experience, or observations, think of examples of cases where individuals acted in ways that exceeded what was expected of them morally. What could you learn from such cases?
4. Some research practices have come into question because of concerns about the correct treatment of the human body, either body parts, or the dead body as a whole. Think about what it means to respect the body, and consider different ways of showing such respect at different times and for different cultural and religious groups.
5. Is genetics or genomics information different from other sorts of information about us? List ways in which it is similar to or different from other forms of information about an individual.

2.11 Coming Up Next

Looking in more detail at how to engage in ethical debates, in a way designed to enhance and hone the skills you already have.

References

Bentham J (1823) An introduction to the principles of morals and legislation. London
Blackburn-Starza A (2011) Genetic Medalling. Bionews (611)
Buchner G (2004) Woyzeck (trans: Farrelly D). Carysfort, Dublin
Dewar S, Boddington P (2004) Returning the Alder Hey report and its reporting: addressing confusions and improving inquiries. J Med Ethics 30:463–469
Dickens C (2009) Hard times. Vintage Classics, London
Editorial (2001) Slicing soup. Nat Biotech 20(7):637
Galton D (2001) Eugenics: the future of human life in the 21st century. Abacus, London
Galton F (1892) Hereditary genius: an inquiry into its laws and consequences. Macmillan, London
Gitschier J (2009) Inferential genotyping of Y chromosomes in Latter-Day Saints founders and comparison to Utah samples in the HapMap project. Am J Hum Genet 84(2):251–258
Habermas J (2003) The future of human nature. Polity, Cambridge

Hanauske-Abel HM (1996) Not a slippery slope of a sudden subversion: German medicine and National Socialism in 1933. Br Med J 313:1453–1463

Homer N, Szelinger S, Redman M, Duggan D, Tembe W, Muehling J, Pearson JV, Stephan DA, Nelson SF, Craig DW (2008) Resolving individuals contributing trace amounts of DNA to highly complex mixtures using high-density SNP genotyping microarrays. PLoS Genet 4(8): e1000167. http://www.hapmap.org

Kant I (1948) Groundwork of the metaphysic of morals. Hutchinson University Library, London

Kelves DJ (1995) In the name of eugenics. Harvard University Press, Cambridge, Massachusetts

Lippman A (1991) Prenatal genetic testing and screening: constructing needs and reinforcing inequities. Am J Law Med 17(1–2):15–50

Marks J (2003) What it means to be 98% chimpanzee: apes, people and their genes. University of California Press, Berkeley

Marteau TM, Richards M (eds) (1996) The troubled helix: social and psychological implications of the new human genetics. Cambridge University Press, Cambridge

McLean B, Elkind P (2004) The smartest guys in the room: the amazing rise and scandalous fall of Enron. Penguin, London

Mill JS (1999) Utilitarianism. Oxford University Press, Oxford

Moore DS (2001) The dependent gene: the fallacy of 'nature vs nurture'. Henry Holt and Company, New York

Muller-Hill B (1998) Murderous science: elimination by scientific selection of Jews, Gypsies and others in Germany, 1933–1945. Cold Spring Harbour Laboratory, New York

Murphy TF (2004) Case studies in biomedical research ethics. MIT, Cambridge, Massachusetts

Nash C (2011) The contradictory place of race in biology. Bionews (613)

Nyholt DR, Yu CE, Visscher PM (2009) On Jim Watson's APOE status: genetic information is hard to hide. Eur J Hum Genet 17(2):147–149

Olson MV (2008) Dr Watson's base pairs. Nature 452(17):819–820

Rapp R (1999) Testing women, testing the fetus: the social impact of amniocentesis in America. Routledge, New York

Risch N, Burchard E, Ziv E, Tang H (2002) Categorisation of humans in biomedical research: genes, race and disease. Genome Biol 3 (7):comment2007.2001–2007.2012

Rothman BK (1988) The tentative pregnancy: prenatal diagnosis and the future of motherhood. Pandora, London

Royal Liverpool Children's Inquiry (2001) Report/The Royal Liverpool Children's Inquiry. The Stationary Office, London

Sartre J-P (1973) Existentialism and humanism (trans: Mairet P). Eyre Methuen, London

Speer A (1970) Inside the Third Reich. Weidenfeld and Nicolson, London

The International HapMap Consortium Hap Map Consent Forms. http://hapmap.ncbi.nlm.nih.gov/consent.html.en

Trials of war criminals before the Nuremberg military tribunals under Control Council law no. 10, Vol. 2 (1949). US Government Printing Office, Washington, DC

Vollmann J, Winau R (1996) Informed consent in human experimentation before the Nuremberg code. Br Med J 313:1445–1447

Weindling P (1996) Human guinea pigs and the ethics of experimentation: the BMJ's correspondent at the Nuremberg medical trial. Br Med J 313:1467–1470

Wells HG (2006) The war of the worlds. Penguin, London

Wilkie AOM (2001) Genetic prediction: what are the limits? Stud Hist Philos Biol Biol Sci 32 (4):619–633

WMA Declaration of Helsinki – Ethical principles for medical research involving human subjects (1964, updated 2008). Helsinki

Chapter 3
What Ethics Is, What It Is Not, and How We Are Going to Proceed

Abstract This chapter gives an introductory account of ethics which at the same time starts to discuss particular issues that arise in genomics research ethics. The difference between ethics and the more focussed concerns and the methods of research ethics committees is explained in detail. The danger of confining ethical considerations to a committee or to one part of a project, including to the area of ethical legal and social implications of genomics is discussed, as are possible problems that might arise in the efforts to raise standards. Features of ethics discussed include: the link of ethics with universality, moral relativism, the aims of ethics and human well-being, the practical goals of ethics, moral motivation, the scope of ethics and precision in ethical argument, and the relation between ethics and other values especially the value of knowledge. A brief account is given of both reasoning in ethics and the relation between ethics and empirical work.

3.1 What Ethics Is Not: Clearing Away Confusion

First of all, this chapter is not intended as a general and inclusive overview of ethics. There are very many such books, and as can readily be appreciated, even an introduction to the topic is more than enough to fill a whole book. My aim is an overview designed to give us enough to proceed in the remainder of the book, and to clear up some common misapprehensions that may hinder our further discussions. This chapter then is designed to discuss features of ethics that must be borne in mind as we work on addressing the various ethical questions of later chapters.

3.1.1 The Business of Research Ethics Committees

At first glance, it may seem obvious that the job of research ethics committees, or RECs, is to do ethics. (I refer to the various committees that are often known as

P. Boddington, *Ethical Challenges in Genomics Research*,
DOI 10.1007/978-3-642-23699-0_3, © Springer-Verlag Berlin Heidelberg 2012

Institutional Review Boards or IRBs in the United States and elsewhere.) This is a natural assumption – natural but wrong. There is a significant difference between the ethics that goes in within the business of the REC and what is meant by ethics in this book. Ethics is not equivalent to the work of the REC, for many reasons.

RECs generally function by applying rules and regulations that are given to them, by law, by international guidelines and regulations, and by their local or national institutions and professional bodies, to the particular cases brought before them. These rules and regulations in their turn rest upon various ethical values. Very occasionally, ethics committees are set up in ways so that their remit includes an input into the formulation of rules. (I have had the privilege of taking part in such a committee for the Australian Department of Health.) However, this is rarely the case, although in other roles, members of RECs may have valuable insights upon which to draw when rules or approaches to the ethical regulation of research are being revised or reconsidered. Very often, too, RECs will look at a project at its commencement, and give what is known as "ethical clearance" for it to continue. Sometimes there is a process of monitoring throughout the life of a research project, but this may involve little more than placing the onus on researchers to report back at various intervals any problems that have occurred.

How is this mode of operation different from the broader notion of ethics with which I am concerned? Let me count the ways.

3.1.1.1 RECs as Advocates for the Subjects of Research

RECs are primarily charged to protect the interests of the subjects of research. This remit gives them an important ethical function, but it is ethics oriented to this particular group. It is true that ethics committees are also concerned not to block valuable research unnecessarily (or at least, they should be so concerned) but this goal of protecting research subjects gives RECs an element of advocacy: a special relationship towards a certain group. Ethics, however, more broadly conceived, is usually considered to have a universal application; that is, any ethical consideration must take into account all who may be affected, and this will include the researchers and anyone else involved in the practice of research as well as any who may benefit from the findings and their future application. This point is not to say that RECs do not also consider such things in weighing up their decisions. For example, there is a clause in the Declaration of Helsinki that stipulates that there may be occasions when archived data may be used without re-obtaining the consent of the original subjects (WMA Declaration of Helsinki – Ethical Principles for Medical Research Involving Human Subjects 1964, updated 2008). The point is rather that in drawing on these considerations, RECs have a licence, as it were, to be more on the side of the research subjects than on the side of anyone else.

Now there are good reasons for this given that research may involve costs for the subjects, and especially given the history of the abuse of particularly vulnerable groups in research. But it is necessary to distinguish between an ethical stance, which to takes a more universal view of balancing the interests of all, and the pronouncements of an ethics committee, which has a special role in advocating for the interests of the subjects.

Ethics and Universalism

On many ethical theories, it is essential that ethical considerations apply universally, although the ways in which different ethical theories do this may vary. For Immanuel Kant, it was essential that when we reason morally, we have to ask whether our actions could be applied universally in all similar situations (Kant 1948). Others have held similar views, including such diverse thinkers as R.M. Hare (1972) and J-P. Sartre (1973).

But who or what is included in this sphere of concern? The utilitarian Jeremy Bentham famously said "the question is not, Can they *reason*? nor, Can they *talk*? but, Can they *suffer*?" (Bentham 1823). All creatures capable of experiencing pain and pleasure are therefore included. However, other philosophers have extended the sphere of moral concern only to humans – or at least to rational beings (so various intelligent aliens may then count; there is also dispute about the inclusion of the intelligent great apes in our sphere of moral concern).

Contrast these positions with the views of a philosopher such as Aristotle (1999). He recognised the claims of personal life on us, but as a result, his sphere of moral concern encompassed those closest to him: friends, family, fellow citizens of Athens, and, only in a vestigial sense, included others from further afield.

Many have argued that the moral progress of a society can be measured at least in part by the inclusiveness of its morality. The abolition of slavery, the extension of political rights to the working poor and to women, and the ending of regimes of apartheid, can all be seen as growing awareness of the universality of moral claims.

3.1.1.2 Ethics Draws on a Wider Range of Issues than Do Many RECs

RECs not only tend to focus on advocacy for a particular group but, by their institutional set-up whereby particular concrete rules and regulations must be referred to, also may address a particular remit of ethical concerns and values. Ethics conceived more broadly has no such limit imposed on it from the outside (although it is a genuine philosophical question where the boundaries of ethical concern lie, for example, whether particular concerns for personal friendships are within or outside the sphere of ethics).

Of course, there are various bodies that address ethical issues more broadly than the typical REC. These include governmental and official bodies and groups of academic researchers as well as organisations advocating for certain groups and/or advocating for, or critical of, certain positions. Some groups with particular concerns in relation to genomics include Genewatch UK, the Genetic Alliance, the Human Genetics Commission, the Progress Educational Trust, the Centre for Genetics and Society, and the Foundation for Genomics and Population Health.

3.1.1.3 RECs, Rules, and Ethics

As mentioned above, RECs are generally confined by the rules and regulations within which they operate, unless they have been set up to have some greater leeway [and indeed different RECs may operate to rather different rules (Glasziou and Chalmers 2004)]; ethics itself has no such limitation. Indeed, especially given the changing environment of genomics research, a major role of ethics is to cast its net wider than local rules and regulations, and critically engage with them; this in fact can be seen to be the major point of this book. Hence, we will be looking at the regulations that exist for research ethics, but asking whether they are fit for purpose in genomics research.

3.1.1.4 RECs as "Hurdles"

Following on from this, it is worth mentioning that this role of advocacy, no matter how justifiable, is certainly a key aspect of why many researchers view getting REC approval as a "hurdle" to surmount, because it is possible to fit it into a model where the interests of researchers are pitted "against" the requirements of the RECs. This may most clearly be seen in cases where ethics requirements are held to be tangible to the detriment of patients (Roberts et al. 2011).

The notion of RECs as barriers to research, and more than this, their actual functioning to this end, can produce a practice whereby researchers may "shop around" for an REC that will give them the clearance they require (Pearn 1995). Often, if researchers wish to recruit subjects from a certain institution or a certain area, then they are restricted as to which REC(s) they must apply. However, it is also sometimes possible for researchers to pick their subjects, whether recruiting new subjects or using already collected samples and data, with reference to the regulations and attitudes of different RECs. Now, is this a good or a bad thing? Much as always will depend upon the individual case. But, in general, shopping around would be poor practice if it was a deliberate attempt to find a "lowest common denominator" REC (metaphorically the equivalent perhaps of getting ethics clearance cheap from "some bloke you met in a pub"); however, if it is actually the case that RECs are putting up unreasonable, unwieldy, or overly lengthy obstacles to ethically sound research, such practice may be not only understandable but also ethically valid if it allows valuable research to go ahead.

3.1.1.5 A Note on Ethical "Clearance" by RECs

RECs function not only to advocate for the research subjects but also, as we have seen, to act primarily at the commencement of a project. Ethical "clearance" is given for the research to go ahead. However, if we are considering ethics more broadly, we can quickly see that there is really no such thing as general ethics "clearance" to any sustained or complex activity. One reason for this is that there are a range of unforeseen possibilities that may be encountered. Another reason is that, even if nothing in the situation changes or is unanticipated, new ethical insights may be had at any time, whether these are spontaneous reconsiderations of one's own or observations and challenges from outside. Many have pointed out these issues in relation to RECs and suggested that ethical clearance should be seen more as a process than as a one-off "go-ahead – the coast is clear" (Mascalzoni et al. 2008). This is to be welcomed, but still illustrates a major difference between the functioning of RECs and ethics conceived more generally, which has a much wider area of concern than the granting of "yeah or nay" to a project.

It is naturally reasonable to consider that if a project has REC approval, then, other things being equal, it is fine to go ahead with it. Although it does precisely give grounds for assuming that what one is doing is ethically appropriate, this can never be an excuse for shutting down one's own critical faculties. We have already seen some examples of how researchers have gone beyond what was required of them, and taken steps to ensure rigorous application of ethical norms. What is crucial is that the institutional and regulatory framework of ethics in research (or elsewhere of course) does not actually act as a hindrance to ethical thought and action. This is such an important point that it will be dealt with at greater length in Chap. 7.

3.1.1.6 Ethics and RECs: A Summary

Painted in their worst light, RECs may seem to be narrowly rule-bound machines for advocating on behalf of research, concerned to put up hoops and hurdles in advance which researchers have to "get through" in order to do research. However, there are of course ethics committees served by dedicated members who conscientiously perform a useful role and which exceed this description. Furthermore, there is much thought given to ways of improving the role and functioning of ethics committees.

Nonetheless, even on an account of an ideally functioning ethics committee, questions of ethics are far wider than the questions that concern ethics committees.

3.1.2 ELSI and Ethics

If our present concern with ethics is to be clearly distinguished from the narrower remit of the research ethics committee, then could it perhaps be identified with the larger aims of the "ELSI" programmes of work, which look more broadly at the ethical, legal, and social implications of developments in genetics and genomics? One might think that the ethics with which we are concerned is actually narrower than "ELSI", since it leaves out the legal and social aspects. However, ethics cannot be parcelled off very neatly. There are many concerns of overlap between these three areas, and in particular much of this book will be concerned with the social context of our ethical judgements.

But a word of caution is needed. Giving this caution paints a picture of how ELSI work might look, at its very worst. This is the stock in trade of cautions: to warn that any unknown stranger might be a bogeyman, when actually he may be a kindly lost uncle from Australia with a large cheque and a box full of games. There lurks a possible danger that just as one might use clearance from an REC to do one's ethical thinking for one, so too researchers and institutions might be tempted to use an ELSI programme to "do" the ethics, that is, to be seen to be "doing" the ethics. But ethics is not something that can be hived off to a sub-contractor.

Ethics is not and should never be "tick boxing": "hey, look at us guys, we've done the ethics! We've got an ELSI programme!" (It's cheap, too, compared to real science – a bargain!) This is not to say that there cannot legitimately be some specialisation. It is rather to insist that if ethics is to be done at all, it is to be taken seriously, and that this means there must be some means, by which it is woven into the warp and weft of the life of a research project and all those involved. The duty and responsibility for ascertaining the quality of the consent rests upon each individual who initiates, directs, or engages in the experiment. It is a personal duty and responsibility which may not be delegated to another with impunity. Nuremberg Code (Trials of War Criminals before the Nuremberg Military Tribunals under Control Council Law No. 10, Vol. 2 1949).

3.1.3 Ethics Aims to Improve Standards . . . .?

It is worth looking carefully at a potential trap, that of assuming that the more regulations, the better; the tighter the constraints, the better; and that this all makes for higher standards. This is not true, or at least not necessarily true. But this damaging assumption is common and on occasion has a discernible effect upon ethical debate. One danger lurks in the hubris of imagining that one's own standards are higher than those of others. But it is vital to note that in ethics, "strict" does not necessarily equate to "better". Often, there will be conflict with other values. For example, raising levels of security checks may impact upon privacy and may be counter to values of efficiency. If values are in conflict, a rise in one standard might represent a fall, or total defeat, for another.

Assuming that where ethics regulation is concerned, the bigger the better is also a trap that may present itself where ethical debate is understood as primarily about protecting one group of people; greater and greater protections for this group or groups may then carry the appearance of greater ethical sensitivity, whilst departing further and further from a fuller appreciation of values. These greater protections may be all for the good, so far as they go. But they may also confuse matters; detract from other values; pass the buck to others; add meaningless empty or unworkable requirements; or otherwise mistake hoops hurdles, and obstacle courses with more enlightened approaches.

There will be disputes about particular cases of course. But there are many examples within the ethical oversight of research where one must question if anything at all is to be gained. For example, in some of my work I have wished to talk to scientific colleagues about their work. This seems a normal thing to do. In fact it is hard to see how it was anything other than a part of my job. But since this might to be construed as "research" into their opinions, I had to fill in various forms to go through the protocol of proving that talking to my colleagues about matters centrally concerned with my job title did not require ethics clearance. Is this improving ethical practice? Or is this just a bit silly, in a country that allegedly values free and open academic discussion amongst equals?

There are many alleged examples where ethics committees apparently work on the assumption that placing tighter restrictions on research is in itself to the benefit of the participants. For instance, consent requirements for research may be commonly limited to "this disease only" or to "this range of conditions only". It may be that this is done under the assumption that this gives participants better information and that therefore better informed consent. Or it may be operating on the assumption that otherwise a slippery slope of research uses might arise. There are many cases where it is a moot point whether or not producing more exacting or detailed requirements actually leads to any improvement, or where the degree of specification can create a distracting illusion of rigour.

It is an open question whether such strict circumscription always produces valuable results. Consider an example. Where consent is tightly circumscribed, one consequence may be that subjects may then be re-contacted for further consent (at some trouble and expense); or that researchers may simply go to other collections where there are more liberal standards of consent (e.g., with permission to use samples and data for any medical research). This is a course often open to researchers because of the wide-scale collaboration and sharing of samples seen in genomics research, which means that there may well be other suitable collections available.

This can have implications that merit some consideration. In some instances, this can mean that samples and data are obtained from populations in other areas of the world; on occasion it can even mean that routes to obtaining participants then mirrors historical routes of colonisation and exploitation of the developing world by the developed, especially where standards are "stricter" in developed countries. Now, in itself, this may be just fine. The point to be made here is that whether or not this is ethically satisfactory needs to be considered; raising standards often has costs attached, and the implications may not always be immediately apparent.

There are also many instances where calls for tighter ethical regulation have led to hand-waving statements that sound terribly marvellous but which do not necessarily bear close examination. There are many such examples, which often involve citing ideals such as "autonomy", "freedoms", "responsibility", etc. These lofty pronouncements may involve drawing on ideals that it is hard to gainsay, but which may be hard or impossible to implement in practice, hard to articulate in a concrete setting, and which may involve conflict with other ideals if taken to extremes. Calls for stricter regulation may also involve passing on responsibilities to others. But this may or may not be appropriate, useful, or fair. We will examine such issues in detail later.

A broad conclusion: We need to be slightly wary of the trappings of ethics, of what looks good, of what calls itself good, and examine it closely to see if it actually is good.

3.1.4 A Note About the Practical Goals of Ethics

One of the principal goals of addressing ethical questions is to solve practical problems, to decide how to act. But ethics is not simply a problem-solving enterprise. There can be value in looking closely at ethics and other issues of value not because we are aiming to "solve" a problem or win an argument, but simply because we wish to have greater self-understanding. It can help us see why and how we hold the values we do hold. Sometimes, as a result of this, we might wish to revise these values. But even if we don't, we may have gained much self-insight. We also may be better equipped to address problems that lie ahead.

Furthermore, and very importantly, ethics is not also simply about "winning" an argument as the only end. Realistically, difference of opinion will always be with us. But the more we understand each other, the more we can appreciate these viewpoints, and perhaps realise why things are not as we ourselves would wish if we ruled the world. This might help us reach compromises; it might lead us to find genuine common ground, to develop creative solutions. At the very least, it might help us cope more easily with disagreement. If you understand and respect the integrity of those whose opinions differ to yours, those different opinions may be much easier to live with.

3.2 What Ethics Is

Giving an uncontroversial characterisation of ethics is, as you might guess, just not going to be possible. The subject of metaethics deals with broad philosophical questions about ethics, such as whether there is such a thing as "truth" in ethics. These questions are beyond the scope of our current inquiry. Normative ethics deals with substantive questions about value and about the rightness or wrongness of our actions. For example, should we try to bring about the best consequences when we act? Or should we concern ourselves with the types of action we are performing, be it an action of taking human life, or of lying, or of inflicting pain, and so on? Other philosophers have argued that an important consideration is the moral character or motives of the person who is acting.

However, it is broadly true to say that the aims of ethics are closely connected with human well-being, however that is characterised, although many philosophers extend the remit beyond our species. Ethics then is broad ranging in its scope. Ethical judgements are also usually held to be in some way generalisable to all relevantly similar circumstances. In other words, if you treat cases differently, you must be able to point to some relevant difference between them that justifies this. Another way of putting this general point is that ethics cannot be partisan: it cannot just apply to people you like, or people you feel some particular and personal connexion to. This point immediately becomes more complex, because there are those who argue that we do have particular reasons for especial concern for those with particular relationships to us, such as our own children, our own patients, and so on – that these personal relationships are themselves an arena of morality. So some people argue that a parent has reason for extra concern about the education of their own child, compared with the education of someone else's child. But a common point is that an explanation must be provided to ground such differential treatment: an explanation that is itself generalisable, so that anyone would have reason to give particular preference to their own children. Otherwise, this claim is not a moral claim.

What Ethics Aims At: Human Well-Being

There have been many attempts to characterise what morality aims at. Many such attempts amount to theories about what might variously be called well-being, or human flourishing. Some philosophers such as utilitarians have considered that quite simply we should aim for happiness (or pleasure) and the absence of unhappiness (or of pain). Others have had a richer account of human flourishing: Aristotle's term "*eudaimonia*" is often translated as "the good life" and encompassed many things, including the development of the virtues and also the possession of friends and other requirements of life.

Important differences exist between those who consider that there is some relatively objective way of characterising human well-being, for example, often based upon an account of human nature – of what we are really like. Others have been sceptical of this, and often then argue that it is up to the individual to set his or her own course in life. Many accounts hold that some notion of autonomy – self-governance – is essential, either on the grounds that only the individual can know what is in his or her interests or on the grounds that the yearning for such liberty of thought and action is an essential part of our natures.

There are crucial differences in accounts of human well-being too between those who see humans essentially as individuals first, and members of society second, and those who consider that it is a central part of our nature to be interconnected with others.

There is a great deal of variation in how ethical principles and concepts are "expressed" in different contexts, depending on what other factors, including other ethical principles, are also applicable. A plausible view is that some ethical principles carry more weight than others and therefore apply more generally in different situations, whereas others are more context dependent.

To take a vivid example, a victim of the earthquake in Haiti in 2010 survived for eleven days under the rubble by drinking cola and beer and eating cookies from the hotel shop where he worked (Carroli and Gilmore 2010). It would be quite preposterous to think of this as wrongdoing, whereas in other circumstances his actions might be described as "stealing from his employer". Of course, he was right to keep himself alive in this way.

There are other actions which it is hard to see could ever be ethically justified. For example, over a period of many years around London, a man broke into the homes of scores of elderly men and women, terrorising all and raping many (Dodd and Hill 2011). I need say more – these brutal crimes can stand as exemplars of ethically vicious actions under all conceivable circumstances.

3.2.1 Universal Ethics or Relativistic Ethics?

A major question in ethics that follows from the idea that it is intended to apply generally, yet noticing the vast variation in ethical codes around the world, is whether or not it can be truly universal in scope or whether some form of relativism holds. Ethical relativism holds that different moral rules and judgements might apply in different societies. A reasonably common view is that there are certain moral judgements that have universal, or pretty near-universal, scope, such as rules against taking the life of another human without some weighty justification, but others which vary from society to society. A common expression of this is found in the claim that it is wrong to interfere with the ethical views and practices of other societies where different ethical concerns apply. A common variant of a relativist position is that even if there is a broad universal morality, the application of moral rules and values within different contexts may lead to differences of detail. So, to take a simple example, if a goal of morality is to avoid harm, an action may have harmful consequences in one situation but not in another. If a goal of morality is to show good will to our neighbours, what counts as good will in one society may vary from what counts as good will in another.

These debates have of course concerned philosophers down the centuries, and it is quite outside the scope of this book to do anything more than to mention them. However, they of course have a deep relevance to our concerns, especially since we are dealing with medical research which involves international collaboration, international data gathering, and is of broad concern to human welfare in general. It is indeed a major challenge to develop international guidelines and principles. Ethical relativism broadly speaking holds that moral judgements are relative to a particular society. But if we are all interacting globally – as we are in many arenas including that of genomics research – then that society is pretty big, and pretty diverse. We better find a way of living in it. Debate and discussion in ethics are precisely aiming to do that.

3.2.2 Thinking Right and Doing Right: Moral Motivation

So ethics is concerned with broad questions about what the right thing to do is; but there is another issue to which we must attend. There are two distinct main problems in ethics: knowing what to do; and *actually doing it*.

Sometimes, we genuinely find it hard to know what the best thing to do is. Should we tell someone just before an important job interview or her final exam that her father has unexpectedly died – or should we wait an hour until she comes out? Should we be "tough on crime" or will this make matters worse? On many occasions, however, we know exactly what to do – we just don't do it. Or we say we'll do it – after lunch. At this point I will leave the reader silently to call to mind a few examples of your own.

Philosophers – and others – have grappled with such problems for centuries: the problem of moral motivation. Some instances might be due to weakness of will (what the Greek philosophers called "*acrasia*"), when we know the best course of action but do not follow it. In many other cases, we are in the grip of self-deception (Boddington 1998). We may manage to distort our beliefs to fit our own convenience. Such cases are also a ubiquitous part of the human condition but can be difficult to spot, especially of course in oneself. Works of literature and religion have concerned themselves with such phenomena and additionally much work in psychology demonstrates the entrenched habits of mind that lead us to distort the world into one easier to inhabit: for instance, see the considerable body of work on the phenomenon of cognitive dissonance (Festinger 1962). Some of the irrationalities to which we are prey may be simply the result of shortcomings in our reasoning, whereas other irrationalities may be motivated. Motivated or not, irrationalities can be very problematic: from an ethical point of view, it is the ones motivated by self-interest and by partisanship that are of most concern.

Work in applied ethics has concentrated on the first of these questions, on what should be done, although the second is at least as important. Hindrances to moral motivation will be included in the discussions of this book.

3.2.3 What Can We Hope for in Ethics?

Given these warnings about the range of opinion on what ethics is all about and what it should achieve, how can we get anywhere? One way of answering this genuinely difficult question is to point out the inescapability of ethical questions: to fail to address these issues is *ipso facto* to take some kind of stand, because whatever we do, we act, or we fail to act – both of these can have moral implications. So we cannot abandon ethics just because answers may be hard to substantiate.

A second approach to answering to the question is to think about whether we are asking too much of ethics:

> Our subject matter will be adequate if we make things perspicuous enough to accord with the subject matter, for we would not seek the same degree of exactness in all sorts of arguments alike, any more than in the products of different crafts. *Nicomachean Ethics* 1094b
>
> Aristotle 1999

In ethics, we can sometimes be certain that particular things are wrong: someone who is already wealthy cheating a blameless elderly man out of his life savings, for instance. It is often easier to be more confident, and more precise, with prohibitions on behaviour such as killing, and promise breaking. However, even these can cause difficulties in application on occasion. When it comes to other questions, we may know roughly what to do, but flounder for precise answers. When does kindness become indulgence? When have I given enough to charity? We often have to be content with less than precise answers, as Aristotle wisely noted nearly 2,500 years ago.

If this seems like a cheat, then think about different areas of science. In some, such as in many areas of physics, great precision can be given and is expected. But in others, this is not only impossible but to chase it is foolish as well. To try to measure someone's blood pressure beyond the limits of accuracy of the instruments is a waste of time; to measure IQ to two decimal places is not only a waste of time but also may give a misleading impression of the power of IQ tests. The same applies to many questions in ethics. Realising this too can also be helpful in coming to workable compromises – we can often think we would make a different precise judgement to someone else, but can recognise that they are in the same moral ballpark, so to speak. That level of agreement may be all we can get, and may be quite enough for us to be able to live with the results.

3.2.4 Ethics and Other Values: The Value of Knowledge and the Value of Scientific Progress

> And so they tell us that Anaxagoras answered a man who was raising problems of this sort and asking why one should choose rather to be born than not – 'for the sake of viewing the heavens and the whole order of the universe'. *Eudemian Ethics*, 1216 a11
>
> Aristotle 2011

I described ethics earlier as being concerned with what might broadly be called "well-being" – however that is understood and characterised. An additional question is what other normative values we have, and how they relate to ethical values. For example, we have aesthetic values, and may value aspects of the natural world such as species diversity or the grandeur of the Horse Head Nebula without considering these to be ethical values. We may value etiquette and politeness without considering these values to be ethical (perhaps because they are insufficiently important, perhaps because they are more obviously parochial to time and place).

A question arises then of what happens when there is a conflict between ethical values and other values. Some hold that ethical values should always take precedence, but this is not necessarily a straightforward issue. For example, you might pit the ethical value of non-maleficence (not causing harm) against aesthetic values, if you consider it could be justified to push someone out of the way, causing minor injury, in order to put out a fire at the Louvre saving thousands of priceless artworks. (Of course, if you did not say sorry afterwards, that would be rather mean.) In fact, of course fire fighters often do face such risks to save valuable property, even though it may be quite right that risks faced to save property should be less than those undertaken to save human life.

One way of understanding such questions contrasts ethical values with other values and attempts to balance them. Another approach may try to find something common to all our values. For instance, one might start to tell a story whereby ethics aimed at human flourishing which could not reasonably reach its full potential without some aesthetic, scientific, and cultural values.

Of presiding concern to the subject matter of this book are any potential conflicts between ethics and epistemic value, meaning the value of knowledge, and in particular scientific knowledge. We have already looked at how any risks of research to human subjects are balanced against the benefits of the research. It is useful to distinguish between two different ways of valuing the knowledge that research might produce. Knowledge might be valued purely *instrumentally*: for what good it brings about. In this case, this might be for any expected medical or economic benefits. Or knowledge might be valued in and of itself: *intrinsically*. Much of the knowledge that we value I would argue falls into this category; human curiosity impels us to find out many of the things that fascinate us with no thought of any further benefit or ulterior reward. It is not just God who has “an inordinate fondness for beetles”; some mere mortals do as well.

We shall return to this topic at various places throughout the book, especially in Chap. 10 on sharing genomic data where it will be vital to ask why we value knowledge in order to answer questions about the ethics of sharing that knowledge.

Epistemology and the Value of Knowledge

Epistemology, or theory of knowledge, is a branch of philosophical inquiry that looks at broad questions about the nature and justification of knowledge. What is knowledge? How do we know what we claim to know? What distinguishes knowledge from mere belief? What makes our beliefs true? Why do we value knowledge?

There are at least two questions to address. One is why we prefer truth to falsity. A second is why we prefer more truth to less truth – that is, the basic

(continued)

impetus behind scientific (and other) inquiry which seeks to increase our knowledge.

All these questions in epistemology have puzzled philosophers since at least the time of Socrates (Plato 1956), and this last question, "why do we value knowledge", is both highly relevant to the concerns of this book, and particularly perplexing. It seems utterly obvious that we do and should value knowledge, but oddly hard to say anything very satisfactory about why.

One answer might be to argue that we value knowledge instrumentally, for the good it can bring us over and above misinformation: it can help us achieve our other goals.

Alternative answers attempt to claim that knowledge has intrinsic (or inherent) value. But can anything more be said about this inherent value? Difficulties in articulating this value are common to many attempts to explain any basic values: in ethics, it is very intuitively plausible to say that it is wrong to hurt an innocent person, but providing an explanation for this beyond "because it hurts them" may be impossible. Likewise, it may be impossible to provide an account of why we value knowledge other than "because it's true", and why we value the process of inquiry other than "because it's interesting", making appeal to "natural" human curiosity.

Conversely, perhaps the pursuit of knowledge has its hazards. This after all is at the root of tales such as the damnation of Faust, the scientist who made a pact with the devil, or of Prometheus, who tried to steal the fire from the gods and was condemned to an eternal punishment. In 1947, Ludwig Wittgenstein wrote:

> It isn't absurd, e.g., to believe that the age of science and technology is the beginning of the end for humanity; that the idea of great progress is a delusion, along with the idea that the truth will ultimately be known; that there is nothing good or desirable about scientific knowledge and that mankind, in seeking it, is falling into a trap. It is by no means obvious that this is not how things are. (Wittgenstein 1980)

3.2.5 *The Case of the "Missing Ethics"*

In science, it is a familiar phenomenon to search repeatedly for something and not to find it. Sometimes, this is because we are not looking in the right place, sometimes it is because we are not asking the right questions, and sometimes it is because there is actually nothing there to see. We have seen how some have suggested that there is too great an emphasis on ethical issues in genetics and genomics: implying that we are looking sometimes for something that is not there. But, conversely, there are also instances where ethical issues have vanished from view but perhaps ought to be resurrected; or where they appear to vanish but are present, yet detectable only by implication or by what is left unsaid.

It is important to watch out for the eclipsing of ethical concerns. This happens in many contexts, for example, it is sometimes the case when people start to talk about the economy, or efficiency, or being practical. It is also important to bear in mind that economic issues are at root ethical issues as well, in so far as they relate to the distribution of health resources and other fundamental needs; as are concerns about efficiency and practicalities.

There is a good example of how ethical issues have been squeezed backstage in the UK's House of Lords *Report on Genomic Medicine* (House of Lords Select Committee on Science and Technology 2009). This report was put together after a consultation exercise in which a number of questions were set concerning developments in genomics. The dominant interests of the report are quite clearly how best to advance the science and technology of genomics, and how to harvest the economic benefits of this for the UK. Within this framework, few ethical considerations were addressed directly, but there are very many of potential relevance here – not least, the question of whether attention to economic advantage might lead us to overlook issues of ethical relevance.

There are many such examples where ethical issues are implicit. Of course, this does not mean to say that we must always and at all times directly address these – there are other things of practical, scientific, and economic import to consider and ethical issues may be in many instances rightly put to one side. However, this is a call to watch out for instances where ethical issues might be unjustly *forgotten*.

3.2.6 Reasoning in Ethics: A Very Brief Summary

Before we begin, a note. The whole place of reasoning in ethics has recently had some bad publicity at the hands of various empirical experiments. For example, it has been claimed that research in psychology and in neuroscience gives grounds for believing that our ethical judgements stem from our emotions, not from reasons (Haidt 2001). However, such research need not put us off our aims, for a number of reasons. For one thing, this science is at a very early stage, and if we do know one thing about the human brain and human behaviour, it is that it is very complex and very hard to understand, and we are liable to have to reconsider any progress we think we have made. Second, these empirical experiments only tell us how people *do* think and act, not how they *should* think and act (and actually, many of them only tell us a limited amount about how people make simulated decisions within an artificial environment). After all, one of the basic premises for concern about ethics is the empirical observation that people tend to skew things too much in their own self-interest. If people are acting "too much" on emotion, then this could be a useful finding for improving the ways in which we reason in ethics. It is common knowledge that people misinterpret statistics, for example, that they allow gut reactions to colour their views of risks of cancer. No one thinks this means that statistics is wrong, although knowing that people have emotional responses to some statistics is useful knowledge the better to help to educate and inform people.

Perhaps, more importantly, such experiments are open to multiple interpretations, and are often interpreted against a misunderstanding of ethics. Claims that the whole basis of ethics has been undermined by "discovering" that it is "really" based on emotion not on reason overlook the fact that moral philosophers have long thought that emotion had a key place in ethics – in fact, some consider it has a central place, for example, "sentimentalists" such as David Hume who famously claimed that "reason is, and ought only to be, the slave of the passions", and considered that ethics has no basis in fact or in reason but rather is based entirely upon emotion or "sentiment" (Hume 1957). Even then, reasoning does have a certain place: facts make a difference. Feelings of righteous indignation towards someone who did you a wrong ought to, and probably will, change when you realise you have got the wrong guy. Even more importantly, a stark division between reason and emotion can itself be seriously questioned. Emotions themselves are not simple "brute" reactions to events but are informed by reasoning and thinking.

Here, we will do no more than provide a very brief survey of the place of reasoning in ethics. We can start from the point discussed above that in acting or making judgements we need to give an account of ourselves in ethics; we need to point to relevant considerations to justify our actions. Given the wide variation in ethical views, there will be a wide variation in what these relevant considerations are. But precisely because there is such wide variation in views, rather than stopping debate with a hopeless shrug of "oh, you think that, I think this", the reverse often happens: we try all the harder to find justificatory grounds for our views (or to search for understanding and common grounds with those who hold different views). One reason for this goes to the heart of the common claim that ethics is in some sense universal – a claim about the meaning of ethics. It seems built into the notion of what it is to be an ethical value that it is not simply a matter of personal taste or very local practice. Practically, too, it is often because we need to live together and come to some general agreement or at least accommodation when there are disparate views, and so again, here is where argument begins.

Arguments in ethics can then often revolve around the question of what is a morally relevant consideration to bring to the table of discussion. Ethical arguments also often revolve around claims of consistency or inconsistency (Glover 1977). If someone is inconsistent in their arguments, or across different cases, we want them to change their views to achieve consistency, or give a reason that explains why the inconsistency is only apparent. They might, for example, try to show why they do accept a particular principle, but why it does not apply in these particular circumstances. In doing so, particular features of the circumstances may be referred to; often, too, the precise definitions of terms will be looked at. Clarity and precision in the concepts that are being used are vital. All these steps will be illustrated throughout this book as we proceed with our arguments.

Reflective Equilibrium

The notion of reflective equilibrium has been highly influential in moral theory since it was introduced by the philosopher John Rawls in his book which is widely considered to be the most influential work of political philosophy of the twentieth century, *A Theory of Justice* (Rawls 1972). It shows a way of advancing our thinking in ethics (and, indeed, in other areas of inquiry) when there is a gulf between ethical theory and practice.

THEORY (GENERAL PRINCIPLES)
|||||||
PRACTICE (PARTICULAR EXAMPLES)

If we find such a gap, what should we do? We could say our theory is wrong – or we could say our practice is wrong. Rawls' suggestion is that we need to move backwards and forwards between theory and practice until we come to a satisfactory accommodation – a balance where no further adjustments are needed, arrived at by a process of reflection. This will involve both theoretical work in ethics and observations. This notion has been extremely influential in work in ethics and more generally in applying theory to practice.

A Conundrum? Reflective Equilibrium and Cognitive Dissonance

Rawls' account requires that when faced with inconsistency between our beliefs and practices, we make a choice until we reach a balanced state of consistency. Unfortunately, much can influence such choices. One of the best corroborated theories in social psychology is that of cognitive dissonance, first proposed by Leon Festinger fifty years ago in his book *A Theory of Cognitive Dissonance* (1962). When there is dissonance in our beliefs, or between our beliefs and our actions, this creates an internal tension which is uncomfortable and which we seek to resolve. So far, so good: this mirrors the situation Rawls envisages. However, ways of resolving this tension are often quite irrational, and often we seek the easy way out. For example, faced with evidence that we may have made a poor decision, we will avoid further evidence that corroborates this, and selectively seek out evidence that confirms our decision and our beliefs – for example, adverts for the car we have just bought. In doing so, we can remain comfortable with ourselves, and it is possible to sustain such zones of comfort with relative ease, especially if we are surrounded by those who do similar.

Evidence for cognitive dissonance provides empirical confirmation of a well-known range of phenomena with which philosophers and others have grappled for centuries of self-deception and other irrationalities that may interfere with moral judgement and action.

(continued)

How do we avoid this? The first step is to be aware of this possibility. The second step is to be ruthlessly self-critical, including with reference to our own vested interests; a third step is to enlist the aid and opinion of others in reaching ethical and other decisions, especially those others who are likely to be critical of one's views.

3.2.7 Empirical Work in Ethics

There is a growing move within applied ethics, in particular with bioethics, to do what is called "empirical ethics". Just as the discussion so far should have made clear, there are many ways in which empirical findings might be of relevance to ethics. The consequences of applying a certain ethical principle in particular circumstances, for instance, could be an important empirical finding. Close attention to what things are of concern or of difficulty to people involved in certain ethical dilemmas likewise could be of great value. Attention to the history of thought in an area is a quite different kind of empirical data that can be extremely edifying, as is work looking closely at the detail of policies. However, where ethics is concerned we need more than purely descriptive accounts; the empirical questions we ask need to be informed by clear thinking about what kinds of empirical data might be relevant, and how they might be relevant.

For example, it is reasonably common to find surveys of what the general public think about a certain issue. Such surveys always give rise to important questions of methodology, such as whether or not they give a representative sample. But a further question is: what would the relevance of the findings be? Suppose that the public were found to be concerned about the implications of genomics research for questions of privacy. What now? It may be, for instance, that these concerns are based partly upon exaggerated media scare stories. But even if such fears are well founded, what should a policy maker do in response? The answer to this is not clear because such concerns have to be balanced against other morally relevant considerations, and often more needs to be known about the nature of the public's views and upon what they are based. (Indeed, one vital question when considering "the public" is "which public" – there are many different points of view amongst those who constitute "the public".) There might be practical ways of addressing these concerns, or they might relate to deeper value questions which may raise more difficult ethical problems. Even if these views are to be taken seriously, an all-things-considered response might be in the end that more important values outweigh them.

For example, people are often asked what their preferences are in certain contexts, see, for example, Kaufman et al. (2008). In some ethical theories, our aim in ethics is to try to maximise the satisfaction of preferences – but this means that we must look at everyone's preferences, not just those of the group surveyed. Furthermore, on many approaches to ethics, the response to data about preferences is quite rightly "so what?" A research subject might prefer to receive full feedback of all findings no matter how irrelevant to their health, and might prefer to get payment for participation. But these preferences are far from all we need to consider.

3.3 Conclusions

Stating exactly what ethics is will always be controversial, but nonetheless a broad characterisation of it can be given. In this book, it will be understood as an inquiry addressing important questions about the welfare of all, widely construed.

As important and useful as the work of Research Ethics Committees might be, it is important to distinguish their remit from ethics in a broader sense.

Questions in ethics often focus on the issue of what to do; but the question of moral motivation is at least as important, although often overlooked. This book will go on to pay attention to both these issues.

Additionally, ethics is often construed as concerned with human well-being or welfare; but there are other values that we need to consider, for example, the value of knowledge and of information. This too will be a focus for our inquiries in this book.

3.4 Study Questions

1. List the ways in which Research Ethics Committees may be concerned with ethics in a narrower sense than is included in the general scope of ethics.
   If you have experience of working on such a committee, or of submitting applications to it, consider if there are any general ethical issues that were not included in the remit of the committee. Of course, you may consider that the particular committees you have worked on take a wider view.
2. Ethical regulations can be important for encouraging improved practice. But can you think of any ways in which increasing ethical regulation may inadvertently lead to deleterious consequences?
3. There are two major obstacles to ethical action: uncertainty about what to do and failure to act. Think of some examples where one of other of these applies.
   Where individuals know what to do but fail to act, what stops them from acting? These cases differ greatly so it may be important to think through several different cases.
4. One way of characterising ethics is that it must be impartial between all those involved. What might it mean to be impartial in ethical decision making?
   How could such impartiality be achieved?
   What obstacles to impartiality might there be in practice?
   Can one act as an advocate for research subjects and retain ethical impartiality?

5. Why do we value knowledge? Think of some particular examples of knowledge that we value. For each of these examples, is the knowledge this valued for its own sake, for what it can do for us, or for both of these reasons?

3.5 Coming Up Next

This chapter has endeavoured to give an overview of ethics, including a consideration of how reasoning, argument and debate in ethics might proceed. We have also so far discussed various ethical principles and concepts. Many books on ethics work by introducing general principles or concepts, and then reasoning from there: a 'big picture' approach. But another approach to developing skills in ethical debate is by just starting to do it, and one way to do this is by critical reading of texts about ethics.

References

Aristotle (1999) Nicomachean ethics (trans: Irwin T), 2nd edn. Hackett Publishing Company Inc., Indianapolis
Aristotle (2011) The Eudemian ethics (trans: Kenny A). Oxford University Press, Oxford
Bentham J (1823) An introduction to the principles of morals and legislation. London
Boddington P (1998) Self-deception, vol 4, Encyclopedia of applied ethics. Academic, San Diego
Carroli R, Gilmore I (2010) Miracle found as rescue teams ordered to stand down. The Guardian
Centre for Genetics and Society. http://www.geneticsandsociety.org/
Dodd V, Hill A (2011) 'Night stalker' sexually assaulted elderly for 10 years after police blunder. The Guardian
Festinger L (1962) A theory of cognitive dissonance. Stanford University Press, Stanford, CA
Foundation for genomics and population health. http://www.phgfoundation.org/
Genetic Alliance. http://www.geneticalliance.org/
Genewatch UK. http://www.genewatch.org
Glasziou P, Chalmers I (2004) Ethics review roulette: what can we learn? Br Med J 328:121–122
Glover J (1977) Causing death and saving lives. Penguin, Harmondworth
Haidt J (2001) The emotional dog and its rational tail: a social intuitionist approach to moral judgment. Psychol Rev 108(4):814–834. doi:10.1037/0033-295x.108.4.814
Hare RM (1972) The language of morals. Clarendon, Oxford
House of Lords Select Committee on Science and Technology (2009) Genomic medicine. The Stationary Office, London
Human Genetics Commission. http://www.hgc.gov.uk/
Hume D (1957) An inquiry concerning the principles of morals. Bobbs Merril, Indianapolis
Kant I (1948) Groundwork of the metaphysic of morals. Hutchinson University Library, London
Kaufman D, Murphy J, Scott J, Hudson K (2008) Subjects matter: a survey of public opinions about a large genetic cohort study. Genet Med 10(11):831–839
Mascalzoni D, Hicks A, Pramstaller P, Wjst M (2008) Informed consent in the genomics era. PLoS Med 5(9):e192
Pearn J (1995) Publication: an ethical imperative. BMJ 310(6990):1313–1315
Plato (1956) Protagoras and meno (trans: Guthrie WKC). Penguin, Harmondsworth
Progress Educational Trust. http://www.progress.org.uk/home

Rawls J (1972) A theory of justice. Oxford University Press, London
Roberts I, Prieto-Merino D, Shakur H, Chalmers I, Nicholl J (2011) Effect of consent rituals on mortality in emergency care research. Lancet 377(9771):1071–1072. doi:10.1016/s0140-6736(11)60317-6
Sartre J-P (1973) Existentialism and humanism (trans: Mairet P). Eyre Methuen, London
Wittgenstein L (1980) Culture and value (trans: Winch P). Basil Blackwell, Oxford
WMA Declaration of Helsinki – Ethical Principles for Medical Research Involving Human Subjects (1964, updated 2008). Helsinki

Chapter 4
Starting Analysis in Ethics: A Practical Guide

Abstract This chapter aims to give further guidance on critical reasoning and analysis in ethics, starting by looking closely at the critical reading of texts dealing with ethics. The aim is to equip readers to begin or to enhance their critical inquiries into ethics. Some basic strategies of critical analysis are illustrated, including crucial questions about the role of authority in a text; the arguments and concepts used; the conclusions drawn; and what assumptions, hidden or explicit, are made. There is an extended discussion of the framing of texts and the importance of spotting frames to critical ethical inquiry. The common frames in genomics of speed and of future promise are discussed.

4.1 Introduction: How to Use This Chapter

We have looked in outline at reasoning in ethics. Ethical principles and ethical concepts have been introduced, as well as the need to look in detail at argument, at any principles, and at any concepts we use. I have also talked about the different elements that need to be drawn together to make moral decisions, and about how this book aims to harness and develop the skills and insights of its readers to this end.

How should we proceed in ethics? One strategy could be called a "top down" approach: a simple version involves taking an ethical theory which provides an overarching framework such as a set of principles, or a procedure for answering ethical questions, and applying it to ethical questions. Many texts in applied ethics take such an approach. In contrast, a more "bottom up" strategy can proceed by looking closely at questions that arise in critically appraising ethical arguments and statements. A really good way of doing this is by paying particular attention to what we can discern by close reading of texts that involve ethical issues (whether or not the ethical dimensions of a text are clearly apparent or merely implied). These strategies are not mutually exclusive, can be used in conjunction, and indeed one can lead to the other: a grounded, bottom-up strategy will reveal broad theoretical questions and assumptions and may help us clarify if there are any overall strategies

DOI 10.1007/978-3-642-23699-0_4,

or principles we wish to follow. There are many advantages to taking a grounded bottom-up approach especially for the beginner: for one thing, it allows you to start asking substantial and useful questions in ethics without committing to any particular theoretical approach. This chapter thus aims to equip readers with some guidelines to such grounded ethical analysis; some of this will be through looking closely at how to critically read and analyse texts in ethics.

For whom is this chapter intended? Some of this chapter discusses rather basic skills in approaching analysis in ethics. This may be useful for readers who come from a background where they chiefly work in other areas such as pure science or for interested lay people. For others, this may seem too introductory. But the chapter will also introduce concepts such as the notion of authority in ethics as well as the idea of framing issues, a vitally important aspect of clear thinking in ethics. The chapter leads into fuller discussion of moral reasoning, and into concrete issues affecting genetics. Readers who feel more confident about reading and analysing texts in ethics are advised then to skim through the chapter, looking at the conclusions drawn about reasoning argument and assessment in ethics. The chapter is written so make it easy to skim and to pick out general points.

4.2 Critical Reading in Ethics: Some Strategies

There is rightly a drive to clarity of expression in writing; however, this does not mean that any good text therefore can be skimmed through and gutted quickly. *There is often much to be gained by reading and re-reading texts.* In philosophy, I was brought up to assume it was natural to expect to have to read key texts more than once; each time you can get something new out of a good piece of writing. Whilst not everything that is written merits such careful inspection, bear in mind that the second time you read something in ethics, you may well find something you had not previously pondered on. This is equivalent to the process of scrutinising data in science, where careful attention to results and to what is observed often reveals insights that were missed the first time around, with gains from further reflection on what is to be looked for, at how the data is to be "read".

In reading, you need to keep all your critical faculties open of course; but I have generally found that it is also often a good strategy at the same time to try to think *from the point of view of the authors*: asking what would be the strongest statement of their position, what would motivate them to reach these understandings. Sometimes it is easier to think "well, that's wrong, and that's wrong, and that's wrong, the guy's a fool", than to think what is right with a text which with you have serious disagreements. Something can be wrong for interesting reasons, and even our opponents can teach us much. Importantly, *the way in which a text reaches a conclusion is just as important as its conclusions*, if not more so.

Differences between disciplines can greatly affect the written text. In a subject like ethics, writing is often much more discursive than it is in empirical scientific disciplines. In science, for instance, you would be less likely to find a paper that considered in detail things that the author thought were wrong: in ethics, this often happens, and you often find a review of all sides of a position. You can, if you like, think of the argument and the reasoning in a text in philosophy or ethics as the data and its analysis that needs to be scrutinised, rather than simply a report of the data and its analysis. Take careful note of any questions that arise as you read: these may be useful clues to where you have misunderstood a point, or they may well be indications of shortcomings in the text, or points of substantive disagreement you have with the author.

In science, there are positions which were once held which are no longer seriously debated (the existence of phlogiston, for example). There are other positions which are under live scrutiny and debate and on which there is room for opinion. (Some theories once thought defeated can of course make a comeback.) It is fair to say that in philosophy in general and in ethics in particular, the arenas for such live debate and difference of opinion are more ubiquitous than they are in most areas of science, and this includes clashes about methodology as well as issues of substance. There is thus much more space devoted to thrashing out such dispute in ethics and in philosophy, and consequently a great deal to be gained from close scrutiny of argument and diversity of opinion.

In law, in presenting a case to a judge, a lawyer has a professional obligation to ensure that he or she makes known to the judge any cases or points of law that go against his or her case. There are high penalties for failing to do so. But there are also great advantages to the lawyer in doing this, because such a thorough search prior to appearing in court will enable him or her to argue in advance against opposing arguments. Take a lesson from this: if you have particular views, then the more you can understand arguments against the position that you yourself favour, the more you prepare yourself to defend this position.

However, be warned: do not suppose that the only reason for careful appreciation of the arguments of the opposition is simply to shore up your own views. In fact, there is much empirical evidence (from research on biases in reasoning) that shows that a person is likely selectively to focus on evidence that supports beliefs he or she already has or that are self-serving rather than evidence that challenges this (Myers 2005). Consciously being aware of this, and testing your beliefs in open debate with others, is a way to guard somewhat against this problem. We will look later in the book in some detail at social and institutional influences on our opinions and how this might affect belief and action in ethics.

4.3 Reading an Article on Ethics: Some Preliminaries

Here we work through some different aspects of reading a paper on ethics commenting on points of relevance to understanding and critique.

4.3.1 What Journal?

Bear in mind that the journal can greatly affect how ideas are presented in ethics and the form of any arguments offered. The major science journals tend to carry policy pieces or opinion pieces from major players; many scientific journals may carry articles that are themselves founded on some empirical piece of work related to ethical questions. The journals devoted to ethics and philosophy tend to have more specialist readership and often go into much greater detail in ethical argument. It is probably fair to say that they are not routinely read by scientists, researchers, or practitioners themselves.

A very important practical point that may not be obvious to the uninitiated: the major science journals which carry pieces on ethics usually have somewhat stringent word limits, and often in ethics this can mean that it is a real problem trying to include adequate details of argument. So many of the papers on ethics that appear in the places where scientists are likely to read them are effectively published lacking necessary details of the argumentation even if the conclusions are based on careful reasoning. (This is in fact one of the reasons why I am writing this book.) If any journal editors are reading this, perhaps a policy of allowing online additional material for ethics pieces might help.

4.3.2 The Range of Work in Ethics

A brief perusal of some titles will give a feel for the range of types of article on ethics.

Informing study participants of research results: an ethical imperative (Fernandez et al. 2003).

Considerations and costs of disclosing study findings to research participants (Fernandez et al. 2004).

The emergence of an ethical duty to disclose genetic research results: international perspectives (Knoppers et al. 2006).

Disclosing individual results of clinical research: implications of respect for participants (Shalowitz and Miller 2005).

Ethics of future disclosure of individual risk information in a genetic cohort study: a survey of donor preferences (Matsui et al. 2008).

Research ethics and the challenge of whole-genome sequencing (McGuire et al. 2008).

Issues of consent and feedback in a genetic epidemiology study of women with breast cancer (Richards et al. 2003).

Informing research participants of research results: analysis of Canadian university-based research ethics board policies (MacNeil and Fernandez 2006).

Donors' perceptions of consent to and feedback from biobank research: time to acknowledge diversity? (Hoeyer 2010).

Researchers to return blood samples to the Yanomanö (Couzin-Frankel 2010).

Hereditary Nonpolyposis colorectal cancer family members' perceptions about the duty to inform and health professionals' role in disseminating genetic information (Pentz et al. 2005).

Genetic knowledge and moral responsibility: ambiguity at the interface of genetic research and clinical practice (Pullman and Hodgkinson 2006).

This selection of titles was gathered by me in a very nonscientific way, by quickly browsing through a pile on my desk. It proved to be a great array of really very different approaches to discussing and contributing to ethical debate. There is very specific research, more general research, papers based closely on policy or analysis of policy, papers that clearly present the author(s) viewpoints, in some cases allegedly driven by their investigations, papers which focus more on empirical contributions, and papers that focus more on ethical concepts.

What is the importance of this? Articles on ethics set out to do a number of different things and therefore can be appreciated and critically assessed accordingly.

4.3.3 Authors and Authority

It is worth noting that the word "author" and the word "authority" have the same root. It is important to watch out for being persuaded of a view by the force of the authority of the author. In fact, there is even a name for basing assessment of an argument upon who wrote it: an "*ad hominem*" argument (literally "to the man"). It is a recognised fallacy. As interesting as it is to know what a particular person thinks on an issue, what is of importance is the strength of their arguments. The *ad hominem* fallacy is to be distinguished from the *general strategy* of watching out for a person's potential biases on the basis of what you might know about them; this can be useful to alert you to unspoken assumptions that affect their reasoning.

Is the author of the piece implicitly drawing on their authority, expertise, or position to bolster their argument? If so, how valid are these grounds – for instance, is this person particularly well placed to make these observations or claims? Or is authority in one area used illegitimately in another? Or are someone's views dismissed for the wrong reasons? For example, there have are often commentaries suggesting that certain people's views on certain matters of ethics and policy cannot be trusted because of their religious faith. Is this really the case? It is vitally important to think through carefully whether such claims are justified.

4.3.4 Authors' Disciplinary Backgrounds and Affiliations

This is not intended as an encouragement to prejudice, but its worthwhile noting the institutional affiliation, and if it is apparent, the disciplinary backgrounds of the authors. We have seen how different points of view can be relevant to questions in ethics. Experience of working in a particular field can very usefully inform aspects of an ethical question (that is one of the reasons why ethics benefits from discussion with others). There may also be broad differences in outlook on ethical issues of those from different geographical areas – as also indicated by our very brief discussion of relativism. This often helps you to be alert to possible assumptions that may not be spelt out in the text. For instance, a writer may be making assumptions about the health-care system based upon the system in the country in which he or she is based – and this could have a major influence on the resulting ethical discussion. Cultural background can make quite a big difference to what background assumptions are spelt out and argued for.

Some papers in ethics are written on behalf of an organisation or larger group, such as a statement of conclusions drawn from a meeting, and it is useful to know. Sometimes, you will see an abstract that gives a really helpful account of what the article is about and where the author is coming from:

Patient Autonomy: Evolution of the Doctor–Patient Relationship (Steinhart 2002)

Summary. *In recent years, much has been said about the shifting locus of control in clinical practice. This article discusses the influence on the physician–patient relationship of recent social and political history, patient culture and experience, and advances in medical practice. I suggest that effective partnerships between physicians and patients are possible in a supportive institutional environment in which participants acknowledge and fulfill their roles and responsibilities in shared decision-making. While I recognize that there is considerable variability in practices across cultures and in individual behaviour, the scope of this article is based on literature and my knowledge, which is representative primarily of the North American experience.*

This excellent example speaks for itself. A quick glance at the abstract and then through the paper shows that whilst focussed on one particular geographical area, it deals with issues of quite general concern (not just to haemophilia as the journal it is in might suggest), presenting an historical view of abstract issues.

4.4 Some Basic Questions to Ask as You Are Reading

4.4.1 Where in the Text Can You Find Arguments and Evidence for Positions?

As you read, you need to work out what the aims are, what the conclusion(s) are, and how well those aims are met and how well the conclusions are supported.

There is no general formula for doing this analysis but some general questions can act as a guide. This can also help to indicate ways in which your particular areas of expertise might be brought to bear in a critical assessment.

What are the aims? For example, to propose a particular general position, specific recommendations, to review policies, to address detailed issues or points of argument, etc.?

What types of reasoning are used and what is the form of any arguments used? For example, is there an appeal to principle, a demonstration that concepts should be thought of or applied in certain ways, an appeal to empirical evidence?

In searching the text for support for the conclusion(s) (that is, for the premises of any arguments), look for:

Statements of empirical fact or alleged fact (and ask upon what evidence these claims are based)
Empirical facts assumed (for example, that there is or is not a certain type of health-care system)
Statements of opinion
Statements of values and principles
Implicit use of values and principles
What references are used to support the argument, and ask if they in fact do so

Are there any broad methodological constraints, for instance, is the work a search for consensus or an appeal to look for difference in different cases?

Are generalisations applied well to specific cases? Is the choice of any example pertinent?

How strong are any claims made? For instance, a claim that a conclusion is "demonstrated" or "follows from" or is "shown by": are these claims really fully substantiated?

In science, to demonstrate a correlation is not yet to demonstrate a cause; to demonstrate a cause is not yet to demonstrate that this is the only causal factor. Likewise, think in this way in ethical reasoning. For example, to show that a particular value is of import in a situation is not in itself to show that it is the only value. What else can be brought to bear in addressing the question?

4.4.2 Key Concepts and Terms Used

What key ethical concepts are used, and how are they understood? Are concepts used consistently in the text? Are they applied in appropriate ways? Are they used in the same ways by other authors?

A first important lesson: key concepts may be understood in different ways by different authors. Just as there may be different understandings of what constitutes various quantitative traits and disease states, as well as different ways of measuring and diagnosing these, all the more so with concepts in ethics. Concepts may be used differently in different contexts, or by different authors. This is often a source of misunderstanding and difference of opinion, and it is often a source of bad arguments. It is useful then to ask if there are there concepts that may be open to misinterpretation.

A second important lesson: some concepts are strongly associated with certain values. Be especially careful with certain "buzz" words that strongly imply values that are hard to deny; call these "apple pie" concepts – nobody is against them. "Freedom" is one example. Who is against this? Yet what exactly does it mean? George Orwell was acutely aware of the persuasive powers of language and encapsulated the moveable feast of meaning in "War is peace. Freedom is slavery. Ignorance is strength" (Orwell 1949). When I visited Moscow in the days of the Soviet Union, a philosophy professor I met said to me, "In the West you think you've got freedom. But are you free to turn up to work 2 hours late if you feel like it and to drink three glasses of vodka before breakfast?" The advisability of this level of alcohol consumption is a distraction from the central point: the word "freedom" can mean all things to all people.

Another one to look out for is any talk of rights. In various jurisdictions, people have certain more or less clearly defined legal rights. Many philosophers have also argued that there are certain universal "natural" rights that we all possess. However, commonly the claim that someone has a "right" to something is little more than the claim that they ought to be able to have it. So consider the basis of any rights claims, and watch out for any slippage between clearly circumscribed legal rights laid down by law, and talk of moral rights, and looser talk of rights which is simply shorthand for some kind of moral claim.

There are some words that always sound good. Conversely, there are some that always sound bad; but which may not be. Paternalism is now so unfashionable in so many quarters that the mere whiff of it can be tantamount to an accusation. I have even seen an "accusation" of paternalism towards children, as if this was *ipso facto* wrong. (Think about it.) The big problem with such assumptions, that certain things are good or bad, is that they can get in the way of arguments and of clear thinking.

Much of this book will involve discussion of key concepts commonly applied in different ways and commonly misused and misunderstood. We will look at autonomy, and at the notion of "respect for persons". "Responsibility" is, if anything, an even more troubling word. There are many calls to take responsibility, or suggestions that someone or some organisation is responsible for a situation, which lack clarity or rigour. But calls to take on responsibility are also difficult to deny without sounding, literally, irresponsible. Careful dissection of the use of such key concepts will take us firmly into the realm of ethical theory.

4.4.3 What Words Are Used? Same Phenomenon, Different Descriptions, Different Values

A key way in which we are persuaded or repelled by different views is precise choice of words. There has been much commentary on the use of dehumanising terms, for instance, the use in warfare of derogatory labels for the enemy: the “Geeks”, the “Nips”. Harsh language can be used to denigrate, and, in doing so, to spur to particular actions, to persuade. Often, the more subtle this is, the more effective it is. Likewise, soft language can be used euphemistically to pass off actions that might otherwise be unacceptable. The *Deutsches Ärzteblatt* of July 29 1933 wrote “Since sterilization is the only safe method to prevent the inheritance of mental diseases and serious genetic disorders, the law must be looked upon as an expression of loving care for the coming generations, and as an act of altruism”, discussing a programme of compulsory sterilisation, for which, to add insult to injury, the victims were (doubtless “lovingly”), charged a fee (Hanauske-Abel 1996).

The use of “technical terms” can also be a way of normalising actions (Glover 1999). “Extermination of a life not worth living” became a standard term in medical education in the 1930s Germany; whether or not, to take one example, the contemporary routine use of the term “TOP” (termination of pregnancy) as medical shorthand is seen to fulfil a similar function will depend greatly on one’s views on the practice of abortion. The lesson for reading ethics though is clear: look for the values implied but not stated.

4.4.4 What Conclusions, If Any, Are Drawn?

Some pieces attempt to sketch out the issues and give an overview of a range of opinions; others attempt to argue for a particular position; some are more focussed on pointing out difficulties; and some attempt to offer solutions. Some pieces look at broad issues, while some are more narrowly focussed on specifics.

Some give firm conclusions, while others hesitate. Do not dismiss a paper just because it has hesitant or programmatic conclusions. What is more valuable is a conclusion that is finely tempered to the arguments and to the evidence presented. However, it is irritatingly common in ethics to find articles that make bland hand-waving statements about “the need to carefully and thoroughly consider . . .”, “we must think carefully about the weighty implications . . .” or to make suggestions that bear scant relation to scientific reality. (*Sometimes*, this is because the journal editors insist that the authors “come up with some suggestions for practice”.) These sometimes come clad in the thin disguise of some lofty-sounding preface: “Mankind stands on the cusp of a new dawn in genetics and must reflect . . ..” Do not be fooled. It can be useful to think through how a conclusion could be made more specific or concrete.

At this point, I could fill an entire chapter with examples of such innocuously empty conclusions. In many cases of course such preliminary and tentative endings are entirely reasonable. Here in contrast is an extract from a conclusion of a more specific empirical study: “In conclusion, our study demonstrates that actual research participants in a genetic epidemiology research want reports regarding their genetic information in the case of serious conditions against which effective interventions are available” (Matsui et al. 2008).

However, the *quid pro quo* for this is that the conclusions may be narrower in scope. These conclusions relate to the findings of one study of a sample of recruits to a particular type of research and may or may not generalise. Note, too, the strength of the claim: "our study demonstrates". You will need to use your judgement to decide if the study really does *demonstrate* this, and with what degree of confidence.

Do the conclusions have implications for policy and practice? In which case, you can ask questions such as whether these are appropriate policy for different settings. Or are the conclusions more focussed upon theoretical concerns such as the principles used or application of concepts? These can be just as valuable, especially as many papers simply assume theoretical positions.

4.5 Looking for Assumptions: Presenting Positions and Framing Texts

Analysis of ethical arguments requires close attention to detail. A step back to look critically at a text as a whole, and its surrounding context, is just as important. The importance of how an issue or concept is presented (or "framed") is central to the careful reading of a text and one's response and critique. It can help to guide us to see arguments and positions in certain lights, thereby helping us towards certain conclusions; it can canalise our thinking. It can either open our eyes to new ways of thinking or help to obscure alternative ways of seeing a debate or an issue. It can also be important to understand the arguments presented to us not just on the level of premises, reasons, evidence, and conclusions, but also in terms of rhetoric: how arguments are presented, what persuasive powers are used.

The importance of framing is underlined by work in a range of disciplines and can be traced back to the work of Goffman (1972). Much work within discourse analysis, social psychology, and literary criticism has concerned itself with these and similar issues. For those used to a relatively formal set of rules that governs how scientific articles are structured, and the type of language that is permissible, this may seem like rather a new concept. However, in fact, presenting what we say or write in certain ways is something we naturally do all the time; spotting the ways others shape what they are saying or writing is also a skill we all have to some degree. What is necessary is to hone this skill in spotting how the presentation of a piece of work might be shaping or skewing the reasoning and the conclusions that follow.

We can think of such a phenomenon of framing quite simply: a perfectly ordinary picture can look great once it is put in a good frame; a frame of the wrong colour or design can ruin an otherwise lovely picture. We often introduce what we are about to say in ways designed to obtain the "right" reaction from our audience. To take an everyday example, we often frame the news that we are not going to do what we said we would do with an apology and an account of excusing factors. "So sorry darling, the traffic is dreadful, I'm really hoping. . ."

An introduction to a book, the blurb on the cover, a trailer for a film, or even just a few simple words of introduction can help to orient us to certain aspects of what is to come – and maybe blind us to others. Different blurbs may set us off on different interpretations: "The story of triumph over adversity and evil: the child of a young single mother and an absentee father, orphaned by his mother's brutal shooting in an unprovoked attack and left at the mercy of the elements, a young boy survives against the odds and eventually meets his aloof and distant father." Would you let a 4-year-old see this film? It in fact describes the plot of Bambi.

In fact, I myself used a common framing device in that last sentence. I said "it *in fact* describes the plot of Bambi." Calling something a "fact" can be a way of emphasising its centrality and of eclipsing other ways of looking at things. If something is a fact, it is true, and, therefore, it stakes its claim above rival versions of reality.

We can miss seeing how issues are presented entirely, and sometimes the framing makes this more likely. We may be so accustomed to seeing things a certain way that we do not even notice that things come in particular packaging. We all know that this can be done in ways which are not untruthful, but which go beyond stark presentation of factual information. Scientific writing requirements leave a great deal of scope for presenting things in particular ways. The presentation of results and statistical analyses will be done to highlight what is most important and novel; the research itself will be presented in ways designed to bring out the most interesting results.

Often such standard framing in the presentation of ideas is of small consequence, but it may be crucially important. This is no more so than where value issues and moral, political, policy, or regulatory concerns are at stake.

The framing of a situation can be achieved not just by how it is presented by others but also by our own interpretations which may be biased. In ethics, what we need to do is to consider things through from the point of view of others, especially those involved in a situation, and consider how our own actions might seem to others. This can provide useful feedback, and may alert us to aspects of a situation that were hidden, or may draw links to other issues, "casting new light" on what is before us. This is recognised in the seeking of second opinions in medicine or elsewhere. This change in perspective can be because others help us to understand, or frame, what is going on with reference to different concepts – we may be looking at the same things, but labelling them differently.

This is such a common device that there are many accounts of it: one excellent explanation can be taken from David McNaughton's book, *Moral Vision*, discussing the story of King David and Bathsheba (1988); Bishop Butler also uses the same example (1856). Bathsheba was married to Uriah the Hittite. David has an affair with Bathsheba while Uriah is away in battle; she becomes pregnant, and so in the hope of covering up who the real father is, David calls Uriah back home. The plan fails! Uriah has made a vow not to have sexual relations with his wife until the battle is won. So, David sends Uriah to back the frontline of battle, ordering that the other soldiers abandon him; Uriah is duly killed, and David promptly marries the widowed Bathsheba.

The prophet Nathan tells then David the following story about two men, rich and poor:

> 'The rich man had exceeding many flocks and herds. But the poor man had nothing, save one little ewe lamb, which he had brought and nourished up: and it grew up together with him, and with his children; it did eat of his meat, and drink of his own cup, and lay in his bosom, and was unto him as a daughter. And there came a traveller unto the rich man, and he spared to take of his own flock and of his own herd to dress for the wayfaring man that was come unto him, but took the poor man's lamb, and dressed it for the man that was come to him.' And David's anger was greatly kindled against this man; and he said to Nathan . . . 'the man that hath done this thing shall surely die . . . because he had no pity.' And Nathan said to David, 'Thou art the man'.

Nathan's parallel story makes David see his actions in a different light; he now understands a different way of describing what he did. This is quite a complex account of what might be seen as "framing"; it is no longer for David a crafty plot to get out of a scrape and to "get the girl", it is a callous abuse of power. Nathan's ability to awaken David's moral vision to appreciate how he had framed the story of his own adultery underlines the importance of genuinely open debate and discussion in ethics.

The moral importance of framing our views of a situation and of individuals has been discussed at length by philosophers and is also the subject of a great deal of empirical research in social psychology. To illustrate this with one example: research on how subjects treat anonymous others has shown that their attitude to them, and the resulting treatment given, can be greatly affected by the prompt (or frame) of something as minor as an overheard comment by a stranger. For example, in one experiment where subjects were required to administer punishments to anonymous others in a learning task, those who heard whispered comments that they were a "perceptive, understanding and otherwise humanised group" meted out far more lenient treatment than those who heard no comment, and the worst treatment was meted out by those overhearing remarks to the effect that they were "an animalistic, rotten bunch; this ill treatment in fact escalated significantly over time" (Bandura et al. 1975). Just as alarmingly, the "frame" of a grey lab coat and other trappings of authority have been shown to greatly bias our actions and our moral judgements (as we shall discuss at length in Chaps. 6 and 7) (Milgram 1974).

The way in which examples are drawn up and extrapolated from in ethics in particular, including a consideration of how this affects framing and how a situation is interpreted, has been discussed at length in an extremely valuable book by Tod Chambers, *The Fiction of Bioethics* (1999). In understanding ethics and ethical reasoning, it will be invaluable to think about what exactly is going on in framings and reframings of situations and the characters in them. However, it is enough to be more consciously alert to this phenomenon to see how it might be affecting how you are reading, and thinking about, what others have to say.

It is also important to be aware that issues can be framed in very conscious ways; for instance, literally in their introduction, or by talking about "what this debate is about", "what this whole debate is really about", "what is at stake", and so on. It can also be framed by simple choice of language and concepts, which, by their selection – perhaps at the expense of other words and other concepts – help us to tunnel our vision, or to cast a new light.

It is also extremely important to point out that showing that a certain "frame" is applied does not in itself mean that this is a problem. It can help to clarify and enrich our thought, but it does not mean in itself, that there is anything "wrong" with our thought. If someone from a background with very different expectations exclaimed to you "but you are taking for granted that women and men are equal/life is worth living/health is good/autonomy is a key value/etc.", you may not wish to alter these assumptions; but it may help to articulate them, and to understand why you hold them. This is a key part of what doing philosophy is about – greater self-awareness. Sometimes, this is of value for its own sake.

4.5.1 Common Framings in Genetics and Genomics: Speed and "Future Promise"

It is undoubtedly true that there have been rapid advances in our understanding of genetics and genomics, and in the surrounding technology and techniques used to analyse the data. However, it is also commonplace for papers to be introduced in terms of such rapid advances; it is also commonplace for this to use related terms such as "urgent" or "pressing". This is perhaps particularly common not so much in scientific research papers but in commentary and review papers. In spotting such frames, it is important not to think simply: is this "true", but to note the frame in order to consider what role if any it is performing in how a text is read. Sometimes, this speed is a part of what the text is about. Sometimes it is just an added extra that may create an added anxiety and urgency about the topic discussed.

> New advances in epigenetics research are being reported at an accelerating rate. Epigenetics research holds great promise for elucidating pathways that affect the progression from environmental exposure to phenotypic expression (Rothstein et al. 2009).
>
> We are at a time of unprecedented increase in knowledge within the rapidly evolving fields of medical genetics and genomics (Al Aqeel 2007).
>
> With the completion of the sequence of the human genome, scientific experts have characterized a rapidly approaching future in which genomic risk information might be used by individuals and health care providers ... (McBride et al. 2008).
>
> The recent publication of the first two individual human whole-genome sequences is significant not only because it symbolizes a tremendous stride forward in our technological and scientific ability to understand the genetic basis of disease, but because it provides a glimpse into the future of genomic medicine (McGuire and Cho 2008).
>
> Sophisticated research tools are rapidly transforming information from the human genome into knowledge that holds great promise for improving disease diagnosis and treatment (Renegar et al. 2006).

Such framings are, if anything, even more visible in media reports of research in genetics and genomics.

It is interesting to note how common it is to hail specific achievements as new milestones that mark the speedy journey into the future – even as the previous achievements hailed as significant milestones often fade as we realise that they only served to show us what we did not know. Note also the frequent reference to the "future promise" of genetics and genomics. It is especially important to watch out for such messages which may be almost subliminal in the text and yet which may sway how we think of the associated ethical issues. If there really is such great future promise, then surely it will be worth many of the difficulties and sacrifices along the way, one might suppose. However, especially where there is vagueness and much is unknown, hand-waving references to such as yet largely unfulfilled promises may do little more than cloud our judgements.

A serious point also needs to be made about the conjunction of framings of speed in genetics and genomics research with ethical questions. It is often claimed that because of this speedy progress we need therefore urgently to address various ethical questions, and there may be a rush to provide answers. We have also seen that there is often pressure put on authors to make policy suggestions or ethical recommendations perhaps prematurely. It is vital to distinguish between *urgency* and *importance*. Many things are important, but not necessarily urgent. Much of the claimed urgency in ethics comes precisely from assurances of large "future promise". These assurances may or may not be true, but are often used to ground this claimed urgency. The implication that great benefits are just around the corner can also send the message that there had better not be any obstacles to these benefits. It is also the case that sometimes urgency is created, or increased, by a subtle conflation of *failing to benefit* with actually *causing harm*; anything standing in the way of new benefits becomes by implication an actual harm.

Harm and Benefit

Two key and commonly accepted principles of medical ethics are those of beneficence and of nonmaleficence. Beneficence is creating good for others; bringing about good consequences. Nonmaleficence is avoiding causing harm to others, minimising any bad consequences.

It is often asked why these two principles are separated: does not it all amount to the same thing? Surely causing a certain amount of good is equivalent to preventing an equivalent amount of bad?

Two main approaches can be taken to provide an answer to this. One approach is to deny that the consequences of our actions are all that matters: the rules under which we act, and/or our motives for acting, may be equally or more important. On most moralities based around rules, rules prohibiting causing harm to others tend to have more weight than those enjoining us to do good. For example, rules against stealing from others may be absolute, or at any rate more robust than rules requiring us to give to charity.

One way in which this has been expressed is in terms of the *doctrine of acts and omissions*, which claims that for any bad consequence, a positive act which brings it about is morally worse than an omission which leads to the same result. To fail to benefit someone may not be so bad as to positively harm them. This doctrine is controversial and can be defended or attacked in various forms.

Another approach accepts consequentialism, the view that the outcomes of our actions are the primary or sole criterion of their rightness or wrongness, but points out that there are significant differences in different strategies for pursing those outcomes. There are complex arguments made here that can only be summarised. One concern is distribution: if we could benefit one person greatly by harming another, this may violate notions of fairness and respect for persons. Hence, it may be argued that avoiding harm is (often) more important than providing benefits. Another concern is with the motives of moral agents. Encouraging good motivation will likely have good consequences all round in the future: discouraging agents from harming others may be particularly important for its effects on an agent's moral motivation.

4.6 Conclusions

Careful reading of texts in ethics, or texts with ethical implications, is already the beginning of critical analysis and reasoning in ethics. Texts in ethics can have a number of different aims and may base their argument upon a range of considerations which all need to be carefully examined. The evaluation of arguments must include a careful consideration of the use of key concepts, and of what may be less visible. Being alert to this, and in particular to the ways in which texts are framed, can be of vital importance in critical reasoning about ethics.

4.7 Study Questions

Read this passage carefully. It is the first three paragraphs of an article (omitting references) (McGuire et al. 2010). A full analysis would of course require a thorough reading of the whole piece; this is an exercise only.

Using cadaveric specimens in genome research presents many benefits: investigators can study multiple organs from one individual, including those impossible to collect from living donors (for example, brains), and health risks are minimised as the individual is deceased. Several genomic research initiatives plan to collect cadaveric tissues. For instance, the Genotype-Tissue Expression (GTEx) project, a US National Institutes of Health Roadmap initiative, is seeking insights into the mechanics of gene regulation by identifying variations in gene expression that are highly correlated with genetic variation. Another example is the Allen Institute for Brain Science's human brain atlases, which seek to create a comprehensive three-dimensional map of gene expression in the human brain. However, in many countries,

(continued)

including the United States, there are not adequate ethical guidelines in place for this type of research.

Scandals involving the removal and retention of organs from cadavers in the United Kingdom led to enactment of the Human Tissue Act, which makes it a crime to conduct genome research using tissues from cadavers without consent. In the United States, cadaveric specimens can be obtained and used for most research with little regulatory oversight, no ethical review and without informed consent. Deceased individuals are not human subjects under US federal regulations. Although the Revised Uniform Anatomical Gift Act recommends that states require individual or family permission to retain cadaveric tissues for use in research, permission is not the same as informed consent. When an anatomical gift is made, only the purpose of the gift is specified (for example, transplantation, therapy, research or education); no discussion of the nature of the research or its risks, benefits and alternatives is required.

However, ethically, individuals retain certain interests after death, including an interest in having their bodies treated with respect and in having their ante-mortem wishes upheld. Living individuals whose organs may be used for research post-mortem may also suffer present harm in anticipation that their corpse will be treated in ways that violate their values or beliefs. Also, family members (whether genetically related or not) have an interest in the disposition of a loved one's remains. Finally, the use of cadaveric tissues in genome research raises additional concerns about privacy. Some have argued that deceased individuals can suffer non-experiential harm, such as loss of reputation that may result from a privacy breach after death, and close genetic relatives have privacy interests that deserve protection. Research has shown that DNA is a unique identifier, and in some forensic cases it has been used to identify the biological relatives of a suspect.

1. Identify any value terms used in this extract. Look especially for claims about consequences of actions or policies, and for claims about what is owed to individuals. Consider the whole range of individuals and groups identified and affected by these issues.
2. How is the, as a whole, passage framed? What effect does this have on the overall argument of the passage?
3. How is the second paragraph framed? This paragraph begins with a statement of UK law, and contrasts it with US law. What is the effect of this contrast in the overall argument of the extract?
4. What empirical claims are made in this extract, and what role do they play in the ethical analysis?
5. The extract claims that for cadaveric research, "health risks are minimised as the individual is deceased." Whose health risks are at issue here? Can the deceased suffer any health risks at all?

6. The extract talks about both legal and ethical issues. Identify where legal claims are made or discussed, and where ethical claims are made or discussed. Is there a clear distinction between these? Should there be?
7. Cadavers are not human subjects in US federal regulations. Why might this be thought to be appropriate? What does it mean in law to have the status of a human subject? What arguments might there be for and against giving a cadaver the status of a human subject? Are there any alternative ways of protecting the cadaver?
8. The extract claims "ethically, individuals retain certain interests after death." What might be said for and against this view? Two different post-mortem interests of individuals are identified. Compare and contrast these.
9. In what ways, if any, could an individual suffer a loss of reputation following a privacy breach regarding their genome after death?
10. What is the difference between "permission" and "informed consent"? What information do you think should be included in any informed consent for the use of cadaveric tissue?
11. On what grounds might an individual's family have the right to give or withhold permission to the use of their cadaver for medical research?

4.8 Coming Up Next

Having given a broad introduction to the issues of the book, discussed ethics and ethical reasoning in some detail, I now move on to introduce to some of the major ethical issues arising in research in genomics.

References

Al Aqeel AI (2007) Islamic ethical framework for research into and prevention of genetic diseases. Nat Genet 39(11):1293–1298

Bandura A, Underwood B, Fromson ME (1975) Disinhibition of aggression through the diffusion of responsibility and dehumanisation of victims. J Res Personal 9:253–269

Butler J (1856) Fifteen sermons preached at Rolls Chapel. Longman, Brown, Green and Longmans, London

Chambers T (1999) The fiction of bioethics: cases as literary texts. Routledge, New York

Couzin-Frankel J (2010) Researchers to return blood samples to the Yanomamö. Science 328 (5983):1218

Fernandez CV, Kodish E, Weijer C (2003) Informing study participants of research results: an ethical imperative. IRB Ethics Hum Res 25(3):12–19

Fernandez CV, Skedgel C, Weijer C (2004) Considerations and costs of disclosing study findings to research participants. Can Med Assoc J 170(9):1417–1419. doi:10.1503/cmaj.1031668

Glover J (1999) Humanity: a moral history of the twentieth century. Jonathan Cape, London

Goffman E (1972) Frame analysis: an essay on the organisation of experience. Harvard University Press, Cambridge, MA

Hanauske-Abel HM (1996) Not a slippery slope of a sudden subversion: German medicine and National Socialism in 1933. Br Med J 313:1453–1463

Hoeyer K (2010) Donors' perceptions of consent to and feedback from biobank research: time to acknowledge diversity? Public Health Genomics 13(6):345–352

Knoppers BM, Joly Y, Simard J, Durocher F (2006) The emergence of an ethical duty to disclose genetic research results: international perspectives. Eur J Hum Genet 14(11):1170–1178

MacNeil SD, Fernandez CV (2006) Informing research participants of research results: analysis of Canadian university-based research ethics board policies. J Med Ethics 32:49–54

Matsui K, Lie RK, Kita Y, Ueshima H (2008) Ethics of future disclosure of individual risk information in a genetic cohort study: a survey of donor preferences. J Epidemiol 18 (5):217–244

McBride CM, Alford SH, Reid RJ, Larson EB, Baxevanis AD, Brody LC (2008) Putting science over supposition in the arena of personalized genomics. Nat Genet 40(8):939–942

McGuire ALCT, Cho MK (2008) Research ethics and the challenge of whole-genome sequencing. Nat Rev Genet 9(2):152–156

McGuire AL, Caulfield T, Cho MK (2008) Research ethics and the challenge of whole-genome sequencing. Nat Rev Genet 9(2):152–156

McGuire AL, Majumder MA, Halpern SD, Swindell JS, Yaeger LV, Gibbs RA, Wheeler TM (2010) Taking DNA from the dead. Nat Rev Genet 11(5):318–318

McNaughton D (1988) Moral vision: an introduction to ethics. Basil Blackwell Ltd., Oxford

Milgram S (1974) Obedience to authority. Harper Collins, New York

Myers D (2005) Social psychology, 8th edn. McGraw Hill, New York

Orwell G (1949) Nineteen eighty-four. Secker and Warburg, London

Pentz RD, Peterson SK, Watts B, Vernon SW, Lynch PM, Koehly LM, Gritz ER (2005) Hereditary nonpolyposis colorectal cancer family members' perceptions about the duty to inform and health professionals' role in disseminating genetic information. Genet Test 9(3):261–268

Pullman D, Hodgkinson K (2006) Genetic knowledge and moral responsibility: ambiguity at the interface of genetic research and clinical practice. Clin Genet 69(3):199–203

Renegar G, Webster CJ, Stuerzebecher S, Harty L, Ide SE, Balkite B, Rogalski-Salter TA, Cohen N, Spear BB, Barnes DM, Brazell C (2006) Returning genetic research results to individuals: points to consider. Bioethics 20(1):24–36

Richards MPM, Ponder M, Pharoah P, Everest S, Mackay J (2003) Issues of consent and feedback in a genetic epidemiological study of women with breast cancer. J Med Ethics 29(2):93–96. doi:10.1136/jme.29.2.93

Rothstein MA, Cai Y, Marchant GE (2009) Ethical implications of epigenetics research. Nat Rev Genet 10(4):224–224

Shalowitz DI, Miller FG (2005) Disclosing individual results of clinical research: implications of respect for participants. JAMA 294(6):737–740. doi:10.1001/jama.294.6.737

Steinhart B (2002) Patient autonomy: evolution of the doctor-patient relationship. Haemophilia 8:441–446

Chapter 5
An Overview of Research Ethics and Challenges from Genomics

Abstract This chapter examines how the traditional basis of research ethics with its foundations in clinical research may not fit the case of genomics research. In particular, the risk of informational harm is a relevant feature of much genomics research that contributes towards many of the characteristic ethical issues in genomics research. Issues also arise when research is carried out in large networks and over lengthy time periods. Some particular key issues that present especially difficulties include: informed consent for individuals, families, and populations; withdrawal from research; confidentiality and privacy; feedback of findings; benefit sharing; ancestry and "race"; and issues arising in public health genomics. Parallels with issues in social science research, which may be helpful for the case of genomics research, are outlined, and the relation between technology and ethics is briefly discussed. Problems with encoding ethics in rules and regulations are also introduced.

5.1 Introduction: Some Challenges in Genomics Research

Many of the current difficulties facing research ethics in many fields, and in genomics research in particular, are being actively debated. Many of these debates derive at root from central features of such research: the nature of genomics information, and the organisation of research both being prominent amongst them. Many ethical issues are also driven by developments in technology that produce new possibilities, or which represent increases in power or speed over past capabilities. Here we start to look in more detail at some of the many difficulties that research in genomics presents and the challenges of fitting genomics research into this traditional model.

To clarify, it should be noted that much genomics research takes place in contexts where the research also has a clinical aspect. I am focussing here on the genomics aspect of research, so leave aside any issues that might arise more specifically from the clinical aspect of research.

P. Boddington, *Ethical Challenges in Genomics Research*,
DOI 10.1007/978-3-642-23699-0_5, © Springer-Verlag Berlin Heidelberg 2012

5.1.1 *The Traditional Basis of Research Ethics and Genomics Research*

We have seen how the traditional basis of research ethics was geared towards protecting the participants of clinical research from harm and from exploitation by unscrupulous researchers. You will recall the earlier discussion (Sect. 2.4) of genetic exceptionalism and whether or not, and to what extent, genetics presents us with particular problems not encountered elsewhere. This chapter looks closely at the extent to which, and ways in which, research in genetics and genomics currently does present particular ethical challenges.

I will argue that one key to understanding the nature of the different challenges posed by genomics research is the nature of possible harms of such research. The most pressing possible harms of clinical research are physical risks of the research itself. The most salient possible harms of research into genomics are, however, informational harms: how information generated by the research might be used and what its implications may be for the research subject and others. The nature of this information will be a key to understanding and tackling many of the ethical problems that arise. The practices that have evolved around the use of genomic information in research, such as the sharing and storage of data and samples, the creation of large research networks and possible security breaches, generate many of the specific practical ethical questions.

One corollary of this is that the potential harms that may accrue from information require attention to the wider social setting within which that information is placed. Information is potentially damaging (or beneficial) within particular contexts, and depending on how it is understood and acted upon by others. Thus, the consequences of genomics information release depend greatly upon larger social facts: the nature of a society's insurance and health-care systems and the reliability and trustworthiness of the police and justice systems, relationships within families and communities, for example. The much-vaunted potential consequence of "stigma" is frequently cited, but of course a piece of information is only stigmatising if it is seen as so by others, and if there are deleterious social consequences. Moreover, the immediate social setting of the institutional structures of research networks will also be a prominent focus of attention and will receive further scrutiny later in the book. Let us now look in some more detail at these points.

5.1.2 *Potential Harms of Genomics Research: Informational Harms*

The proximal physical harms of most genomics research are minor, and relate simply to the slight risks of blood collection. In some research, indeed, only a cheek swab is taken. Whilst of course it is still necessary to minimise such risks and to explain possible deleterious effects to subjects, these risks are so minimal that for present purposes we shall put them to one side; these are not ethically insignificant, but nonetheless not our focus.

The risks of genomics research, then, are largely informational, in particular connected with: the extremely *large amount* of information that is generated by genomics research; the *nature and power* of that information; and the *possible consequences of any release* of that information – whether to the person him – or herself, to particular others, or in any possible security breach.

What is it about DNA information that is of particular concern? Much of the data will be of scant interest in itself, but some will of course pertain to traits and disease status or risk, and also crucially to the identification of individuals and to questions of relatedness and ancestry. Although in many disease areas there is little progress in terms of use of genomics research to improve prediction and risk profiling, for example see Paynter et al. (2009), in other disease areas there has been more progress in prediction, and to some extent in treatment, as a result of the application of genomics information (Heard et al. 2010). Examples of "researchers behaving well" in Sect. 2.7 all involved recently highlighted problems arising from the power of DNA data and the increasing ability to interrogate and analyse it. An individual's DNA may pertain to his or her disease risk or status, and to phenotypic information about traits including not only physical but also psychological and behavioural traits. A relatively small amount of DNA data may uniquely identify an individual out of a large population (Lumley and Rice 2010). DNA data may also pertain to relatedness to others, and to ancestry, and hence an individual's DNA can also have significant implications for others who are related to an individual or who share ancestry.

It is also important to realise that false or misleading genomics information may of course be of equal if not greater power.

Genetic Information: Some Implications

Genetic information gained from genomics research may reveal:

Disease status: this could reveal a definite prognosis or a risk of disease

Phenotypic information: such as eye colour, or information relevant to the calculation of quantitative traits

Behavioural traits

Relatedness: for instance, in some studies that collect information from family groups, genomics information may reveal discrepancies in reported relationships – e.g., non-paternity, or information indicating that a sibling is adopted, or perhaps suggestive of a family grouping where an illegitimate grandchild is passed off as a child

Rates of inbreeding including information suggestive of possible incest

Possible ancestry

Implications for family members, such as changing disease risk, or unsettling expectations of relatedness

(continued)

For population groups – may reveal disease susceptibilities, suggest information concerning ancestry, may reveal kinship patterns including, for example, the prevalence of non-paternity within a particular group, or marriage between close relatives. May potentially conflict with a group's own notions of their history

5.1.3 Research Networks

Recall how the Nuremberg Code stated that researchers have a personal obligation to ensure that informed consent to research is obtained, and that this obligation cannot be delegated away to another (3.1.2). The surprise with which this clause has been greeted by some of the researchers with whom I have discussed it does not spring from any general moral dissipation amongst them, but has far more to do with how much medical research, including a great deal of research in genomics, is currently conducted: in large and often international networks with very large numbers of recruits, and often in long-term studies where data may come from samples taken some time ago. Often there is a need to compare geographically diverse populations. Data and samples are routinely shared amongst members of a research team or consortium and also with other research groups. Most subjects taking part have no personal relationship at all with most of the researchers.

Some research in genetics takes place of course on rare diseases, some of which affect only a handful of people worldwide, and some research takes place within a smaller community where there are more personal relationships involved. For example, a patient group may consist of individuals or families who have a long-standing relationship with a hospital attached to a research institution and the research may be carried out by a team which includes clinicians treating the patients. But the large network model is extremely common in genomics research, and indeed the size of such networks and the size of collections of samples from subjects is increasing, largely driven by the spur to increase the power of the data collected.

It is not only genomics research that is carried out in large networks. Much other medical work also is conducted in this way, and this will be increasingly the case with the establishment of biobanks, such as UK Biobank which has recruited 500,000 people. However, it is certainly the case that the scale of such research projects, and the distance between time and space that results between researcher and subject, and between different researchers, poses a great challenge to the ethical regulation and conduct of research. It is, literally, thousands of miles away from the situation envisaged at Nuremberg of a doctor experimenting on subjects known to him or her in his or her own hospital, university, or clinic.

Why Is There a Need for Such a Large Number of Recruits in Genomics Research?

Those familiar with genomics research will be familiar with the often large sample sizes used. Especially when research is looking at common multifactorial diseases, the genetic influences on disease susceptibility may each have only very small effects, and there may be very many of these, together with other influences such as environmental variables. Moreover, different genes may interact producing variable influences within individuals with different genetic makeup. Hence, in order to control for all these multiple interacting variables, and to detect any small effects, it is vital to have large samples sizes to control for all the differences. Even if a genetic variation has only a small impact upon an individual's risk of disease, it may be of great biological importance in uncovering the mechanisms of the disease. Moreover, some important genetic variation may be relatively rare and only discoverable if large samples are used. Additionally, because of gene–gene and gene–environment interactions, the impact of genetic variations on disease susceptibility may vary from population to population. Hence, large and international sample comparisons are often of great scientific importance (Hattersley and McCarthy 2005; Channon and Watkins 2004; Ioannidis et al. 2007).

Some Examples of Large-Scale Research Projects and Collaborations Involving Genomics Research

The Wellcome Trust Case Control Consortium was established in 2005 and now represents a collaboration of about 50 research projects not only across the UK but also including international collaborations, studying the genomics of common and less common diseases such as diabetes types 1 and 2, coronary artery disease, ankylosing spondylitis, pre-eclampsia, and schizophrenia, as well as reading and writing abilities in children and pharmacological reactions to statins. Tens of thousands of cases and thousands of controls are used.

The 1000 Genomes Project aims to take advantage of falling costs of sequencing to sequence the DNA of 1,000 people with the aim of cataloguing all variants in the population studied with frequencies of 1% or more.

The Personal Genomes project aims to recruit 100,000 members of the public openly to share genome sequence and other information with the research community.

Biobanks are being built up around the world to conduct long-term studies on a range of diseases using a range of techniques including but not limited to genomics research. UK Biobank has recruited over 500,000 adults aged

(continued)

between 40 and 69 to study "the diseases of the twenty-first century: including diabetes, heart disease, and dementias".

The Kadoorie Biobank study is a similar project involving 500,000 Chinese recruits.

Many cohort studies of disease include genomics work. For example, the Framingham Heart Study has been underway since 1948 in the town of Framingham, Massachusetts, and is now into the third generation of recruits. This study now includes research into the genetics of heart disease.

The Avon Longitudinal Study of Parents and Children (ALSPAC) recruited (almost) all children born in the county of Avon, UK in 1991. This study is conducting research on many aspects of lifestyle and health and includes genomics research.

5.2 Ethical Issues

In what follows, I intend to give an overview of issues and different responses to them. Following chapters will then look at some of the key issues in greater depth.

5.2.1 Informed Consent: Complexity of Information and Rapid Change

As we have seen, informed consent has become a central plank of research ethics. It presents a number of difficulties (O'Neill 2003) and perhaps especially so in genomics research. One issue concerns how "informed" consent can be, or needs to be, in complex areas of research. A difficulty is simply the sheer complexity of much of the research, and the question that follows of how much a potential recruit needs to know in order to give fully, or adequately, informed consent (Mascalzoni et al. 2008; Panoyan et al. 2008). This is a general problem in medical research, however, and especially for certain individuals or population groups who may lack educational privileges (Chokshi and Kwiatkowski 2005), and it is a moot point whether genomics research is especially complex compared to other research.

There is in practice much debate and a spread of opinion on how much information should be provided to attain adequate informed consent, with some projects giving an outline of the research only; at the other end, there are projects such as The Personal Genome Project which requires subjects to follow a study guide and pass an exam to demonstrate their understanding of genetics and social and ethical issues before recruitment. Such online educational resources as this study guide show effective ways of improving the quality of informed consent (and indeed genetics and genomics boast a wide array of excellent online educational resources).

A particular reason for difficulty concerning informed consent relates not so much to the capacity of recruits to comprehend the complex science of genomics, but is inherent to the nature of research: it advances our knowledge and our understanding. This field is advancing so rapidly that it is often genuinely difficult to foretell to what uses data and samples may be put. The original terms of consent may become sketchy at best or redundant as technology continues to outstrip itself. For example, subjects to research recruited only a few years ago might have got the impression from the consent information that a few points of data within their DNA would be examined; whereas, because the technology developed and became cheaper very rapidly, it is routine at the time of writing for a genome-wide association scan to scrutinise a million points of data for each individual. If at the time of recruitment, both the researchers and the subjects envisaged far less than this, is consent informed? Is it valid? This is a moot point, and the same basic problem applies to a great deal of research now, especially where data and samples are stored over a long period of time for renewed uses.

A suggestion to deal with this is to obtain broad or blanket consent to the broad aims of the research (Lunshof et al. 2008; Hofmann 2009; Hansson et al. 2006). However, critics argue that this fails to provide adequate information for true consent and therefore does not comply with the goal of respecting individual autonomy. Against this, the importance of informed consent may be argued to be lesser in genomics research where potential harms are mostly informational rather than physical (Taylor 2008). However, it is still an open question whether such harms are necessarily less important: what is clear is that informational harms require that we take note of social context: aside from the case of an individual contemplating personal information him or herself, informational harms always involves looking at who else has that information, and how they might interpret and use it.

Blanket or Open Consent and Restricted Consent

The Estonian Genome Project consent form makes provision for consent to an open description of future genetic research: consent is given for samples "the use thereof for genetic research, public health research, and statistical and other purposes in accordance with the law." The consent form states that data will only be available "only in coded form so that the identity of the gene donor remains unknown to the receiver of the data".

The Personal Genomes Project also operates on open consent to future research. However, in this project no guarantees are given of anonymity or privacy, given the advances in the interrogation of data that have made such promises harder to uphold.

Why might it be important to give restricted consent only? The case of the Havasupai and Arizona State University (2010) illustrates some particular

(continued)

issues. Because the case settled rather than proceeding to a full court case, details of what exactly occurred are harder to ascertain. However, the Havasupai's case was that the original consent had been given only to the study of DNA to study diabetes. Publications making use of the tribe's DNA covered many other areas, including mental illness and theories about the tribe's geographical ancestry. These conflicted with the tribe's own version of their history, and fears about the stigma associated with mental illness were also expressed (Mello and Wolf 2010).

5.2.2 Informed Consent: Individuals, Families, and Populations

Informed consent to medical research has standardly been a matter for the individuals concerned. In fact, in many ways that was the whole rationale: the requirement of informed consent operated as a means to give power to individuals to decide what to do. Note that informed consent can be justified in more than one way (Pelias 2006). One model justifies it on the grounds that it is a key part of valuing individual autonomy. Another model justifies it on the grounds of harm avoidance, that individuals are on the whole best suited to make judgements about their own welfare. But there are questions for both approaches. We have seen how the risks of genomics research relate not so much to physical harm, but to risks from the possible uses to which information generated might be put. We have also seen how genomics information may have relevance to related individuals and to population groups. In the light of this, traditional models of individual informed consent have been challenged. More will be said on these topics later in the book when we discuss autonomy in greater depth. Meanwhile an introductory survey to the issues will be given.

In many cases, particularly with research involving isolated populations, or easily identifiable communities, population or community consent may be obtained in addition to individual informed consent (Mascalzoni et al. 2008; Austin 2002). This may be seen to add an extra ethical safeguard in that the community who may be affected by the research findings as a whole is allowed some say in the project. In some instances, this has meant that communities have had some input into how exactly the project was to be carried out. For example, in the HapMap project, communities were consulted on how the population groups to be included in the research were to be named, and care was taken to ensure that questions asked of different groups were sensitive to the local community concerns and standards, for example, avoiding asking questions about ancestry which were deemed insensitive in particular communities (HapMap Consortium 2004). Such a process is not without its difficulties (Beskow et al. 2001; Rotimi et al. 2007). In different communities, different processes of consultation may be appropriate depending on the political and power structures of the community; and there is not necessarily any guarantee that all members of the community will be equally happy with the outcome (Mascalzoni et al. 2009). On a traditional model of research ethics, such community consultation or consent may be seen as an interference with the freedom of individuals to give or withhold consent regardless of the views of their communities. We will consider this issue again shortly in considering parallels with social science research (Sect. 5.3).

It has also been mooted that since genomics research potentially has implications for subjects' genetic relatives, then these should therefore be involved in decisions to participate (McGuire et al. 2008). For example, if there were to be any security issues with the information, this could have potential implications for relatives; in some countries, the courts can obtain information from genomics research studies and this may have implications for relatives, for instance. But giving relatives any sway over an individual's choice to take part in research or not is not only practically difficult to achieve, it is a radical departure from the standard individualised model of research ethics (Boddington 2010). For such reasons, discussions have so far limited themselves to a half-way house of suggesting that potential recruits discuss the research with family members. Nonetheless, it remains the case that the information generated by genomics research may sometimes be of great potential import to those other than the individual participant.

5.2.3 Withdrawal from Research: Difficult, but then, Is It Really Necessary in Genomics?

Traditionally, as explained earlier, research subjects have the right to withdraw from research at any time without penalty. Withdrawal there means physically removing oneself from the study. What does this mean in genomics research? In some projects, continuing involvement may be needed, but where a sample has already been collected, what is it exactly to withdraw? This could refer to removal of one's sample from any continued research, or on some interpretations, removal of one's data.

A practical problem often presents itself with genomics research. Often an individual's sample may be contained on plates with many other samples, and often too, samples are split and shared between different arms of a larger research project meaning that individual samples may be physically located in different places. Hence, there can be logistical problems, and often large implications in terms of time and costs, in withdrawing individual samples. Furthermore, much genomics research consists in analysing data which often may be aggregated. Of course, once data have been publicly released or published, withdrawal of an individual's data is just not possible.

Responses to such difficulties include the suggestion that the option of withdrawal should no longer be offered for genomics research (Eriksson and Helgesson 2005).

Proponents point out that the rationale for withdrawal from genomics research is quite different than for clinical research. If a subject is being administered a drug or other therapeutic regime, then to continue to do this once the subject wishes to withdraw would be a violation of basic rights of bodily control, and in law would constitute an assault. In genomics research, however, the scope of withdrawal is over a sample of DNA or of blood or saliva, or of data generated from this. So, some argue, it is premised upon less ethically pressing grounds.

However, much here depends on views of the nature of the harms involved. Some might argue that the possible informational harms of genomics research should ground rights of withdrawal every bit as strong as those required to protect individuals from the physical (and other) potential harms of clinical research. Moreover, the importance of control over genomics information may be seen as crucial protection of an individual's autonomy: such issues will be further discussed in Chap. 6.

5.2.4 Confidentiality and Privacy in Genomics Research: New Challenges to the Power to Protect Data

Information is confidential to a particular relationship (for example, between doctor and patient, client and counsellor, and priest and parishioner) if it is understood that it is to remain within the bounds of that relationship. Often, the law backs up certain standards of confidentiality, such as that between certain professionals and those who seek their services. Laws such as those regulating medical confidentiality, as well as more specific laws governing the protection of personal data, such as the UK Data Protection Act 1998, provide protection for such confidential information.

Information is private if the individual to whom it pertains has rights of control over that information, such that others would infringe privacy rights were they to violate these. Spheres of privacy may be physical (so that, for example, the police may not search your person or your property without proper authority) or may relate to information, such as rights to privacy of correspondence.

Confidentiality is routinely guaranteed in medical research. One routine way of protecting information about an individual is to anonymise data, and additionally to aggregate data from individuals into larger groups, such as groups of 1,000 individuals, so that data about any one individual is protected. However, a large problem presents itself with data that are especially rich and powerful: it has particular power to identify and also to reveal things about the individual to which it pertains. I have already discussed how developments in science and technology mean that the past assurances about the levels of protection of personal data gathered in genomics research may be difficult to give with such levels of confidence (Botkin 2001; Church et al. 2009; Gitschier 2009; Lowrance and Collins 2007; McGuire and Gibbs 2006). Naturally, steps are being taken to rack up the levels of protection as the potential for breaches in security is created by increasingly sophisticated ways of analysing data. Methods of analysing data might be applied to any of the numerous kinds of detailed information routinely being gathered in individuals – as in data gathered by commercial organisations or from analysing users of social networking sites – but it is often seen that genomics information is especially problematic because so many points of data are held about individuals. A debate is to be had about whether it is any longer viable or responsible to offer full confidentiality in genomics research as previously taken for granted in medical research.

5.2.5 Confidentiality and Privacy in Genomics Research: Whose Confidentiality, Whose Privacy?

Worries about the security of genomics information may be real but they do not affect the fundamental claim that an individual has a right to protection for their genomics data. But genomics data itself present inescapable difficulties here. Genomics data about an individual just do have implications for other individuals, both those in an individual's immediate genetic family and also in relevant population groups. Recall the case of James Watson's apoE gene (Sect. 2.7.3) (Nyholt et al. 2009). Suppose that Watson has an identical twin who was happy to know this and to post it on the Internet? Suppose an individual is happy for his or her DNA data to be posted online: this means that others can then make pretty accurate educated guesses about the genomes of the individual's siblings (Cassa et al. 2008). We will look at such issues in the next chapter when we examine the notion of autonomy. These problems, like many in genomics, relate to the informational harms that might ensue, and like many issues in genomics cast doubt upon an ethic premised upon individuality.

5.2.6 Feedback of Findings in Genomics Research

There is considerable debate about whether or not, and in what circumstances, research results should be fed back to individuals who took part in research. This debate is ongoing in many fields of research, for example, research involving MRI and other imaging. There is increasing agreement that broad general findings of such research should be accessible to research subjects (and to the general public) – the discussion of the value of knowledge in Chap. 10 will look more closely at some related issues. However, there is a more intense debate about the return of individual results. In much research, findings may be of some relevance to the individual clinically, and sometimes there may be "incidental" findings. For instance, an imaging study might discover an otherwise unknown tumour.

This debate raises a general question about DNA (and other) data, which also arises in relation to questions about privacy: does an individual "own" his or her genetic information? Or does an individual have some other rights of control over his or her genetic information?

Feedback of findings in genomics research is discussed in more detail in Chap. 8 as an example of the application of the notion of respect for persons. Particular issues highlighted by genomics research include the power of such research to produce great volumes of information about each person; and particular problems encountered in very large, geographically disparate research networks concerning the transmission of information and pathways of responsibility.

5.2.7 Benefit Sharing in Genomics Research

It has standardly been the case that those taking part in research are clearly told that they themselves will get no benefit from participation. In fact, in some types of research, there may potentially be some individual benefit, for example, the chance to trial new medication that may have some therapeutic benefits to the individual – but it should be made clear that this is not to be expected. This is part of attempts to ensure that there is no inducement to take part in research that might compromise a subject's voluntary participation.

Mostly those taking part in genomics research gain no benefit whatsoever. There can be some variation in even this, however. For example, in some countries recruitment to research is easier than in others. The population of Sweden, for instance, has a reputation for high rates of research participation. In other areas, however, and for different types of research, some small incentive may be needed, and sometimes this may take the form of offering some benefit. For example, in a project where initial clinical measurements such as blood pressure and blood lipids are taken, these may be fed back to the individual or to their general practitioner as a small incentive to participate.

There are two areas where the question of benefits to genetic or genomic research has been prominently discussed. One concerns the sharing of profits from any such research. It is standard practice that even if a commercial arm of a research project might anticipate profits, such as from the development of a therapeutic agent, recruits are told clearly that they themselves will not benefit financially. It has been suggested that trust in the researchers is more important to recruits than financial issues (Weinfurt et al. 2008). But there have been some prominent cases in medical research where this has been challenged (Moore v Regents of the University of California 1990). These relate to the creation of immortal cell lines which have gone on to reap great profits for those involved professionally but not for those from whom the cells were taken. Such cases involve complex issues of property rights, proper reward for effort, and individual cases have their own complexities. The more general issue of the "ownership" of the fruits of research, including the role of public and private ventures in genomics research, will be discussed in more detail when we discuss data sharing in Chap. 10.

There has also been discussion of benefit sharing in research, including in genomics research, where that research involves the participation of particular population groups (Chokshi and Kwiatkowski 2005; Knoppers 2000; Rotimi et al. 2007; Seguin et al. 2008). This relates to the nature of research in genomics – that it is necessary for scientific reasons to include different population groups – as well as to the international structure of much genomics research. Some of the groups that researchers would like to study include those who have by any standards a small share of the earth's riches to call their own. These include various indigenous groups, such as Native Americans, as well as whole populations within the developing world. A particular issue concerns differences in the researcher and the researched; often the research group is situated within the developed world, the researched within the developing. In such cases, greater attention has been placed on what benefits if any should go to those taking part, both the individuals and the communities concerned. Benefits include small rewards for taking part, promises that any drugs produced would be available to the community, and steps to help build local infrastructure. In such cases, historical considerations apply. These relate both generally to the history of colonisation and exploitation; and more particularly to concerns over "genetic piracy" where certain groups have been concerned that their genetic information, or the gene pools of important local species, may be exploited for gain by others.

However, given that a major rationale for excluding significant benefits for research participation was to avoid providing inducements to take part in research, it is therefore necessary to ask if different standards are being applied in different cases, if this matters, and what the most important ethical questions are in different contexts.

5.2.8 Genomics Research, Ancestry and Race

The history of ethics abuses associated with genetics means that it is almost inevitable that the question of genomics research and "race" will arise (Walloo 2006). Indeed, I placed the word "race" in quotation marks precisely because research in genomics itself has done a great deal to undermine the notion that human beings can be divided into distinct and separate groups. There is in fact a great deal of genetic overlap between geographically diverse groups of humans as well as much evidence of interbreeding between different groups. However, it is the case that using statistical generalisations about different population groups, individuals can often be assigned to one group or another, and in some instances claims of ancestry can be verified, or rendered uncertain, by genomics (although the claims of certain commercial companies to provide information about ancestry have been subject to the criticism that they are overstated) (Editorial 2001).

It is also the case that much genomics research routinely takes note of the population group from which individuals originate (Risch 2006). In much research it is essential that population group is taken into account in order to rule out spurious associations, for instance, where the disease or condition of interest has different prevalence rates between different populations, and where therefore other spurious differences might affect findings. In such instances, use of information about a person's ancestry is a methodological tool. In other work, differences in disease prevalence rates and disease severity between population groups are the direct subject of genomics research. Indeed, without research into different population groups, injustice may result, as findings from one group may be misleading if applied to another group where there are relevant genetic differences. For instance, mutations common in one group but not in another may affect reactions to pharmaceuticals; genetic variants that increase risk by a certain amount in one population group may give a different risk in another group (Foster and Sharp 2005).

Concerns that research is "racist" raise a number of highly complex issues which have been, or continue to be, the subject of argument and scrutiny (Haga and Gingsberg 2006; Kahn 2005; Obasogie 2009). Precisely because racism is such a weighty and emotive issue, retaining a clear and rational focus on such issues can be hard. This book does not attempt to delve thoroughly into the debates but simply to introduce some of the main issues. One group of concerns is with how information generated by genomics research might be used to good or to ill. There are concerns about the power of information generated by genomics to produce or to reinforce stigma experienced by certain groups. Since stigma is in the eye of the beholder, misinformation about genetics and genomics is relevant here. So, for example, such information may be used by some to uphold an essentialist or reductionist view of race, which feeds into stigma and continued racism. Other problems might arise when genomics research conflicts with a group's views of their own ancestry (Mello and Wolf 2010). However, genomics information can also be used to link us as well as to separate us, as when we are called upon to consider the common links of genetics.

The Genes That Bind Us: Statistics, Genes, and Ethics

Just as some have tried to argue that science shows differences between different individuals and different groups of people, others have tried to argue that the science of genomics points more to what we have in common. We have noted how nearly all ethical theories have at heart the idea that ethical concern extends universally. This universal concern has been based on different things: for example, on our common rationality, on the notion of personhood, on the idea that we were all created equal and in God's image. The human genome is often spoken of in universal terms, for example as "our common heritage". These ideas can be brought together and the common human genome used to ground a universal ethical concern. Here, focus on what we have in common genetically overrides the genetic differences between individuals and populations:

Genetic knowledge may bring people into a special moral relationship with each other. We have therefore proposed the following concept of genetic solidarity and altruism, which promotes the common good.

We all share the same basic human genome, although there are individual variations which distinguish us from other people. Most of our genetic characteristics will be present in others. This sharing of our genetic constitution not only gives rise to opportunities to help others but it also highlights our common interest in the fruits of medically-based genetic research (Human Genetics Commission 2002).

Others have remarked however on the human need for contact with others which can lead us to find commonality in a wide range of different factors, suggestsing genetic links are just one amongst many.

5.2.9 Can Genomics Information Be Inherently "Bad"?

Indeed, some claims that have been made carry the implication that some genomics information is in and of itself problematic. Such views have been applied to research that might discover certain links of "race" or population group with traits such as those linked to intelligence, or behavioural traits such as aggression; or to the notion of race itself. After all, it's perhaps perfectly possible, scientifically speaking, that we might have discovered that humans could be classified into distinct races. What then? Would this matter? Whilst it is doubtless the case that most would loudly shout that this had no political or moral implications, it is easy to see how others might use this as fodder for exclusionary views. Could such knowledge in itself be bad? Or is it just how it is used? Such debates continue prompted by various concerns. For example much attention has been focussed on the question of whether research into the genetics of sporting ability that is looking for certain genetic variants common in a couple of African populations might be in itself problematic (Blackburn-Starza 2011).

Whether tacit or explicit, acceptance of genetic essentialism is one underlying reason why these debates often take place in relation to genetic and genomics knowledge. The view that information about our genes is fundamentally information about who we are gives that information a particular power to spoil or enhance our identity.

5.2.10 Public Health Genomics

There are a large number of questions in public health genomics arising chiefly from possible applications of genomics knowledge (Avard and Jean 2008;

Burke et al. 2006). Many of these questions then raise issues common in public health ethics, for example, in relation to equity and health disparities in different populations; issues in health economics and resource distribution; the acceptability of various interventions in different cultural and religious communities; issues about the proper realm of government interference in private lives; issues about responsibility for health; and issues relating to reproduction.

For example, many issues arise in relation to the application of genomics research to newborn screening; to population screening for disease or for carrier status; to pregnancy or pre-conception screening, and possible selective termination of pregnancy or selective implantation of embryos, or other reproductive techniques. With the application of genomics knowledge in various contexts, from some quarters come worries about the spectre of eugenics and worries about reproductive ethics.

Another realm of issues relates to applications of genomics knowledge in treatments. At the moment one main potential clinical applications is in pharmacogenomics, where there are hopes that those most likely to benefit from certain pharmacological agents, or most likely to suffer side effects, could be identified (Altman et al. 2011). This includes hopes that previously abandoned drugs could be reinstated if those who will suffer bad side effects could be identified. Possible worries about this include what will happen to those patients who are excluded from available treatments because their genetic profile means they will not benefit or will suffer unacceptable side effects, although a legitimate reply to this is that such would not benefit anyway, and are at least no worse off than before.

The focus of this book is on ethical issues relating more directly to genomics research rather than to its applications and there is simply not enough space to do justice to these extremely important and often complex questions.

Stigma

Genetic information is often said to be linked to the danger of stigmatising individuals or groups.

A case often cited is that of the stigmatisation of carriers of sickle cell anaemia in the USA as a result of a screening programme that took place in the 1970s (Walloo 2006). Under this programme, African Americans, who have a higher risk of the disease than is average in the US population, were screened. The screening process detected both those with the disease and also unaffected carriers. Unfortunately, carriers subsequently experienced problems associated with misunderstandings of their status, such as refusal of employment and insurance. They were thus said to be stigmatised.

This example is often cited as evidence for the stigmatising powers of genetic information, and particular conclusions are often generated from this, typically related to the strengths of protection needed around genetic or genomic information. But in many ways, it is not a very good example, since it involves an already stigmatised group. The ways in which layers of stigma may be added to stigma are generally unexamined. The differential operation of stigma in different settings and for different groups is rarely discussed in the many loose discussions of the stigmatising powers of genetics and genomics. Stigma is often referred to as a "bad" outcome without any real discussion of what exactly it might involve, but precision is necessary to analysing how to respond to it and possibly address it (Kenen and Schmidt 1978).

So what is stigma? The classic work which still retains its status in the field of sociology as an iconic exploration of the concept is Erving Goffman's book *Stigma: Notes on the Management of Spoiled Identity* (1974). An individual who suffers stigma is disqualified from full social acceptance. As used originally by the Greeks, stigma referred to visible bodily signs of moral status, such as marks indicating that the person was a slave or a criminal. It is now used more generally to refer to the associated disgrace of such lowered social status. Features, whether visible or invisible, of the stigmatised individual mark him or her out as socially abnormal, and this "deviance" detracts from being seen as a full human with a whole personality.

The notion of "courtesy stigma" applies to those associated with the stigmatised, such as carers for a person with disabilities, who by association have their social identity likewise affected.

It is important to note that stigma can exist only given the attitudes, beliefs, and behaviours of others. In some societies, certain things will be stigmatised whilst others are not, and what is stigmatised can change over time. For example, the stigma of homosexuality has diminished over the last few decades in many countries, and attempts have been made to address the stigma of those with physical and mental disabilities through, for example, legislation and steps to make public buildings more accessible. Within societies there also will be variation on how the stigmatised are treated: they may be ignored, excluded, treated over-enthusiastically. Some may collude with social stigmatisation and some may actively seek to resist it.

An IVF "mix up" case reported by the BBC illustrates well the separation of stigma from the person or quality itself that is stigmatised. Two children born of IVF sued the clinic in 2011 for using the "wrong" sperm which meant that they had significantly darker skin colour from either of their parents, and in the area where they grew up, they suffered taunts and bullying because of this. In rejecting their case, the judge commented that there was no inherent disadvantage in having any particular skin colour and any fault lay entirely with those responsible for the bullying. (Parents of children with darker skin lose IVF appeal 2011.)

5.3 Ethical Issues in Social Science Research: Parallels and Lessons for Genomics Research

Research in the social sciences also has suffered in many respects from being squeezed into the straightjacket of the requirements of clinical research. It is an inherent part of much social science research that much of social science is reflexive and involves a great deal of thought and critique of its own methodology; and, especially perhaps as a relatively young discipline, it is also self-consciously researching new techniques of gathering data and new methodologies. A close consideration of these matters will be highly instructive in reflection on the current status of genomics research regulation.

One problem concerns methodology. Methodologies in social science are varied, often themselves the subject of controversy even within their own discipline in terms of their validity and effectiveness. So it may be harder to convince a committee that research is based upon a "rigorous scientific basis". Moreover, a checklist of ethically pertinent questions at the start of a research programme may present problems for research during which new issues, questions, and methods may emerge. This inherent unpredictability of much social science research mirrors the unpredictability that flows from rapid changes in science and technology and which characterises much genomics research today. In each case the net effect is similar: problems in forcing research into a particular form, and problems arising from a method of regulation that requires one-off approval at the start.

Issues of consent may be highly complex. Paul Atkinson describes this problem well in relation to consent to ethnographic research, where the ethnographer wishes to negotiate access to observe and participate in a group, where interest in individuals is of them *as members of that particular group* (Atkinson 2009). But informed consent to research is premised entirely on the notion of individual consent. As we have seen, this includes individual powers to withdraw consent at any time and without reason. But for an ethnographer to operate on such a premise is to risk the whole research project, as one change of mind would jeopardise the whole negotiated consent. Atkinson argues that an insistence that rights of withdrawal more suited to a clinical context be maintained in ethnographic research is only likely to incline researchers to fudge the issue.

Likewise, in genomics research, often it is the population group which is of interest, and the population group with whom consent is negotiated. Likewise, there may be great difficulties for the genomics researcher in complying with standards of withdrawal of research participation. Suppose one person started having doubts towards the end of a million dollar project? As we have started to see, and as will be discussed at greater length in the following chapter, assumptions about the individual basis of research ethics, and especially use of the notion of individual autonomy as an ethical principle in research, present serious difficulties.

There are also similarities with the problems discussed of how well informed a research subject needs to be. Atkinson points out that in obtaining "informed" consent to social science research, only a few aspects of the project are generally touched upon. The possibly unanticipated outcomes of a project are left alone; subjects are not routinely asked to consider how they would feel were published findings to indicate that some negative traits were associated with social groups to which they belong; the multiple analyses to which data may be subjected are not discussed. "As a consequence, most 'informed consent' is so minimal as to be virtually worthless" (p. 8). Similar questions can arise in genomics research: how far do you spell out the possible negative effects of being linked to a trait that someone, somewhere, might find stigmatising?

Atkinson questions what harms might realistically ensue from participation in social science research, and whether then the great weight of regulation is really required. Possible harms are mostly social, economic, and psychological, and, as with genomics research, many of these are potential informational harms. At the same time as narrowly focussing on rules purportedly designed to minimise or eliminate harms, other wider social and cultural questions are ignored by the restricted menu of ethics regulation, such as the promotion of social justice, or the situation of marginalised groups. Similar questions are also being asked in the context of genomics research, where there are many ethical issues that go beyond the remit of the ethics committee, for instance, as we have seen, concerning benefit sharing, social justice issues concerning marginalised groups, or the use of genomics research findings in ways that may overlook social causes and hence help to perpetuate social inequality.

One lesson to be learnt from this is that it is not simply from the point of view of genomics that serious questions can be asked about the standard model of clinical research ethics regulation. These problems include: shoehorning research ethics into a fixed check list of issues; regulating research at a point in time before its commencement; assumptions made about research and about human understanding upon which the notion of informed consent rests, but which do not bear much scrutiny; assumptions made about the priority of the individual in how we conceptualise ethical issues.

5.4 A Note on Technology and Ethics

Does technology really present us with new ethical questions and problems, or does it simply present old issues in newly highlighted ways? There are reasons to consider that advances in technology and in scientific understanding can sometimes present us with new ethical questions and dilemmas, and can also shape how we might answer and address such questions.

Sometimes ethical issues arise because technology produces gradual progress until we reach a point where issues become more problematic or more noticeable. So, for instance, humans have long used various substances to chance their mental states. Alcohol, for example, has been brewed for millennia. There is many a courtship that might never have happened were it not for the courage – or loss of inhibitions – gained from a nerve-steeling drink. The use of substances then for what might broadly be called "human enhancement" is not new. Recent advances in pharmacology however have increased our powers to create psychoactive substances, and increased our powers to comprehend what effects they might or might not be having. Questions that are currently being raised about the ethics of human enhancement technologies could thus be seen as simply developments of older questions.

Technologies can however present us with questions which are substantially new, or which we have not considered previously. For instance, the question of when exactly human life ends is one brought into focus only by developments in medical technology which enable us to keep people's organ systems functioning long after they would have otherwise died. The need to develop clinical criteria for death that address such issues was presented to us by technology. With this too comes not just a new question but a question requiring a greater degree of technicality and precision.

Many questions about technology are whether or not the use of new technology is doing more harm than good. For example, it is now possible to save extremely premature infants, or infants born with a range of serious medical problems. Making decisions about when intervention might simply cause greater suffering is a serious matter. It highlights too the general moral question of whether one should always act to save human life. Such questions are not in themselves new. Answering them may require a mixture of reference to ethical values and the use of technology itself to answer factual questions, for example, questions about prognosis. The dilemmas presented so vividly by developments in technology may prompt some to revise their views on ethical values or on the balance to be struck between different ethical values.

Sometimes technology presents us not with discrete questions that it is possible to formulate very accurately but with a cluster of problems, many of which are hard to anticipate, understand, and conceptualise. For example, the development of social networking sites through the Internet raises questions that were hard to anticipate and which are still hard to formulate. These include questions such as how people are forming relationships and learning about the world; about responsibilities for intervening in cases of "cyber-bullying"; what it is to trust in an online world; about online identities, and how these impact upon notions of identity more widely; about the democratisation of power; about rights of access to and control over information; the use of information by third parties, and so on.

Indeed, developments in technology often do not pose new ethical questions so much as provide a spur that reorients our views of who we are, where we are in the world, and how we relate to the rest of the universe. A prime example of this is given by the diverse and often heated reactions to evolutionary theory, reactions which are still going on and indeed which help to form an important strand of thinking in relation to genetics and ethics, insofar as questions about genetic reductionism, and genetics and identity both individual and group, arise and perplex us.

The difficulty of anticipating all the questions that arise with scientific and technological developments is one of the many reasons for needing the close cooperation of those involved at all levels, and of encouraging discussion between these parties, because often only problems become visible in the context of such free rolling communication, and because often it will take the collective wisdom and thought of many to find ways through difficulties.

Often too, the new that is brought to us by technology is not "bad" per se, but it is something that we need to think through and understand if we are to manage it properly. It can unfortunately be the case that simply by discussing something as a problem this can polarise views and encourage the knee-jerk reaction that practices should be banned, or that those who question them must want to do so. The popular press in chasing headlines and controversy is responsible for a great deal of such irritatingly irresponsible debate.

5.5 Conclusions

We have examined how the old skins of the ethical regulation of research may not accommodate the new wine of genomics research. Particular problems that have arisen have demonstrated strains in the framework. Many of the problems relate to information in genomics and any potential harms of this. A focus on genomics information leads to questioning an ethic that is focussed upon the individual for two reasons. One, genomics information pertaining to one person is often intimately relevant to others. Two, ethical issues concerning information largely depend upon its reception by others: often harms (or benefits) of information can only be identified with relation to the social context, to particular others who possess, interpret, and react to, that information. These and other reasons give a strong impetus to thinking through carefully the individualistic framework of traditional research ethics. Much of the remainder of the book will thus be concerned to examine such questions, through looking at the notion of individual autonomy, ideas of respect for persons, and the social and institutional context of research ethics regulation. We also need to consider general difficulties with using a system of rules and regulations for making ethical decisions. Consideration of ethical issues arising from genomics information will be cast further afield when we consider values attached to such information in debates about sharing it.

In addressing these challenges we need to consider too, that many of these issues raise concern beyond the powers both of individual researchers, and of genomics research *per se* – so much will depend on outside forces such as the political system, the economy, the health-care system, and "general social attitudes". The particular temperature of some issues, such as benefit sharing in research with developing nations, is deeply set by lengthy histories that are of course outside the control of researchers in genomics. In some cases, there can be a blurring of responsibilities: for example, how much responsibility do researchers have to ensure that their results are not misinterpreted? More loosely, how much responsibilities do researchers have for general genetics education, for example, for addressing misplaced notions of genetic determinism and reductionism that in turn might feed into the stigmatising of those with genetic disorders or of particular population groups? A further area of investigation therefore will be that of the issue of responsibility in research.

Rules and Ethics: Some Challenges

If we work out what to do, ethically, by reference to a system of rules, then this generates a very big problem: if our rules prohibit something we would like to do, so long as we are in charge, we can just change the rules to fit. Philosophers have grappled with this and related problems for centuries. A system of rules has to have some justified basis for the rules it generates to have any moral bite and to prevent such rule changing.

Any system of rules needs interpretation to account for different situations, and also needs some system of enforcement: whether this is informal, relying on the strength of individual consciences of the effects of social disapprobation, or a system of formal punishments or penalties.

It is easy, and common, for a system of rules to evolve over time in ways that may or may not depart from their original intent.

Rules may not fit reality very well: there may be new developments or points of detail that leave unanswered questions.

In particular, rules may not fit social processes or the behaviour of social actors very well.

The codification of rules into a formal account may be necessary to create a formal system that can be used in different situations but may also act to narrow the focus of the rules.

The rules may be followed in accordance with the strict detail but not in accordance with their spirit, or minor rules which are codified may be followed at the expense of more important values which are not codified. (We even have a word for a person who acts like this – a "jobsworth".)

5.6 Study Questions

1. For research on identifiable populations, it is now often the case that community consultation takes place as part of the consent process. Suppose that a particular identifiable community decided against taking part in a genomics research project. However, an individual male member of that community disagrees and signs up to a research project where his DNA is made freely available, and research findings about ancestry are then made public which conflict with the community's own account of their heritage. Discuss the ethical issues involved.
2. Compare and contrast the reasoning behind allowing individuals freely to withdraw from a clinical trial of a new pharmaceutical, with withdrawal from a genomics research project where a single sample of blood was donated several years previously.

3. Large corporations collect and analyse immense amounts of information about individuals from the use of bank cards, the Internet, etc. From information about shopping habits, for example, analysts are able to infer with considerable accuracy a great deal about an individual's personal characteristics, relationships, and lifestyle. Compare and contrast the privacy implications of such information collection and analysis with concerns about privacy with regard to genomics information. Are there any concerns particular to the use of genomics information?
4. Consider the following case. Some individuals know that they are at risk of developing Huntington's disease and may decide to take a test or not. This is a very serious, untreatable, and distressing disease. Many do decide not to take the test and in all cases, counselling is provided for those who are considering the test. Now suppose the organisers of a genomics research project decided it was ethical to provide feedback of major incidental findings to those taking part, and an individual is found to have the gene for Huntington's disease. Such an individual may have had no idea he or she was at risk, for example, if he or she was adopted or did not know the cause of a parent's death. Should this result be fed back and, if so, how should this be done? What implications do such cases have for how a genomics research project might go about informing recruits that significant individual incidental findings will be fed back?
5. Are there some research questions that should not be asked? There has been a history of controversy surrounding research into race and intelligence. Intelligence is a highly complex matter affected by many variables including diet, social environment, and genetics, and likewise humans cannot be clearly and simply divided into separate "races". However, it is possible that there are some genetic factors related to different aspects of intelligence that may be distributed differently amongst different population groups. Is it wrong to investigate these? If so, why, and what implications might this have for research on various mechanisms of intelligence that may be of importance, for example, in understanding intellectual disabilities?
6. Different population groups take part in an extensive genomics research project which eventually leads to the development of improved screening and treatment for particular genetics conditions. However, there are of course cost implications in implementing screening and treatment. Are there any obligations to ensure that all populations which took part in the research leading to these developments are equally able to access the benefits of the research? If so, whose responsibility is it to ensure that equitable access is possible?

5.7 Coming Up Next

So far, we have looked generally at how ethical issues that arise in genomics research, discussed the nature and scope of ethics, and how ethical reasoning might proceed. In the remainder of this book, we are going to look in more detail at some of the issues central to ethical challenges in genomics. There are too many diverse issues for this book to be able to cover all possible issues thoroughly; moreover, one of the main difficulties is that emerging issues may not have been predicted. The book thus is aiming not to cover all angles – because that would be both impossible and misleading – but rather to point to some major and general issues that will help to equip readers to take these issues further. So, in discussing particular issues, I will be examining some key concepts in ethics: these are all concepts which pose general problems in ethics, often problems which are particularly acute or distinctive within genomics. I will also be developing the discussion of reasoning in ethics by illustrating further methodologies to employ in thinking about ethics. The topic of the next chapter is that of autonomy.

References

1000 Genomes Project. http://www.1000genomes.org/. Accessed July 2011

Altman RB, Kroemer HK, McCarty CA, Ratain MJ, Roden D (2011) Pharmacogenomics: will the promise be fulfilled? Nat Rev Genet 12(1):69–73

Atkinson P (2009) Ethics and ethnography. Contemp Soc Sci 4(1):17–30

Austin MA (2002) Ethical issues in human genome epidemiology: a case study based on the Japanese American Family Study in Seattle, Washington. Am J Epidemiol 155(7):585–592

Avard DGG, Jean MS (2008) Involving the public in public health genomics: a review of guidelines and policy statements. GenEdit 6(1):1–9

Avon Longitudinal Study of Parents and Children. http://www.bristol.ac.uk/alspac/. Accessed July 2011

Beskow LM, Burke W, Merz JF, Barr PA, Terry S, Penchaszadeh VB, Gostin LO, Gwinn M, Khoury MJ (2001) Informed consent for population-based research involving genetics. JAMA 286(18):2315–2321. doi:10.1001/jama.286.18.2315

Biobank UK. http://www.ukbiobank.ac.uk/. Accessed July 2011

Blackburn-Starza A (2011) Genetic medalling. Bionews (611)

Boddington P (2010) Relative responsibilities: is there an obligation to discuss genomics research participation with family members? Public Health Genomics 13(7–8):504–513

Botkin JR (2001) Protecting the privacy of family members in survey and pedigree research. JAMA 285(2):207–211. doi:10.1001/jama.285.2.207

Burke W, Khoury MJ, Stewart A, Zimmern RL (2006) The path from genome-based research to population health: development of an international public health genomics network. Genet Med 8(7):451–458

Cassa CA, Schmidt B, Kohane IS, Mandl KD (2008) My sister's keeper?: genomic research and the identifiability of siblings. BMC Med Genomics 1:32

Channon KM, Watkins H (2004) Coronary artery disease genetics: bigger is better. Eur Heart J 25:900–901

Chokshi DA, Kwiatkowski DP (2005) Ethical challenges of genomic epidemiology in developing countries. Genomics Soc Policy 1(1):1–15

Church G, Heeney C, Hawkins N, de Vries J, Boddington P, Kaye J, Bobrow M, Weir B (2009) Public access to genome-wide data: five views on balancing research with privacy and protection. PLoS Genet 5(10):e1000665

Editorial (2001) Slicing soup. Nat Biotechnol 20(7):637

Eriksson S, Helgesson G (2005) Potential harms, anonymisation, and the right to withdraw consent to biobank research. Eur J Hum Genet

Estonian Genomes Project. http://www.geenivaramu.ee/. Accessed July 2011

Foster MW, Sharp RR (2005) Will investments in biobanks, prospective cohorts, and markers of common patterns of variation benefit other populations for drug response and disease susceptibility gene discovery? Pharmacogenomics J 5(2):75–80

Framingham Heart Study. http://www.framinghamheartstudy.org/. Accessed July 2011

Gitschier J (2009) Inferential genotyping of Y chromosomes in Latter-Day Saints founders and comparison to Utah samples in the HapMap project. Am J Hum Genet 84(2):251–258

Goffman E (1974) Stigma: notes on the management of spoiled identity. J. Aronson, New York

Haga SB, Gingsberg GS (2006) Prescribing BiDil: is it black and white? J Am Coll Cardiol 48 (1):12–14. doi:doi: 10.1016/j.jacc.2006.04.017

Hansson MG, Dillner J, Bartrarn CR, Carlson JA, Helgesson G (2006) Should donors be allowed to give broad consent to future biobank research? Lancet Oncol 7(3):266–269

HapMap Consortium TI (2004) Integrating ethics and science in the International HapMap project. Nat Rev Genet 5:467–475

Hattersley AT, McCarthy MI (2005) What makes a good genetic association study? Lancet 366:1315–1323

Heard E, Tishkoff S, Todd JA, Vidal M, Wagner GP, Wang J, Weigel D, Young R (2010) Ten years of genetics and genomics: what have we achieved and where are we heading? Nat Rev Genet 11(10):723–733

HGC (2002) Inside information: balancing interests in the use of personal genetic data. Human Genetics Commission, London

Hofmann B (2009) Broadening consent and diluting ethics? J Med Ethics 35(2):125–129

Ioannidis JPA, Boffetta P, Little J, O'Brien TR, Uitterlinden AG, Vineis P, Balding DJ, Chokkalingam A, Dolan S, Flanders WD, Higgins JPT, McCarthy MI, McDermott DH, Page GP, Rebbeck TR, Seminara D, Khoury MJ (2007) Assessment of cumulative evidence on genetic associations: interim guidelines. Int J Epidemiol 37:120–132

Kadoorie Biobank Study. http://www.ctsu.ox.ac.uk/kadooriebiobank. Accessed July 2011

Kahn J (2005) Misreading race and genomics after BiDil. Nat Genet 37(7):655–666

Kenen RH, Schmidt RM (1978) Stigmatization of carrier status: social implications of heterozygote genetic screening programmes. Am J Public Health 68:1116–1120

Knoppers BM (2000) Population genetics and benefit sharing. Commun Genet 3:212–214

Lowrance WW, Collins FS (2007) Identifiability in genomic research. Science 317 (5838):600–602. doi:DOI: 10.1126/science.1147699

Lumley T, Rice K (2010) Potential for revealing individual-level information in genome-wide association studies. J Am Med Assoc 303(7):659–660

Lunshof JE, Chadwick R, Vorhaus DB, Church GM (2008) From genetic privacy to open consent. Nat Rev Genet 9(5):406–411

Mascalzoni D, Hicks A, Pramstaller P, Wjst M (2008) Informed consent in the genomics era. PLoS Med 5(9):e192

Mascalzoni D, Janssens ACJW, Stewart A, Pramstaller P, Gyllensten U, Rudan I, van Duijn CM, Wilson JF, Campbell H, Quillan RM (2009) Comparison of participant information and informed consent forms of five European studies in genetic isolated populations. Eur J Hum Genet 18 (3):296–302. doi:http://www.nature.com/ejhg/journal/v18/n3/suppinfo/ejhg2009155s1.html

McGuire AL, Caulfield T, Cho MK (2008) Research ethics and the challenge of whole-genome sequencing. Nat Rev Genet 9(2):152–156

McGuire AL, Gibbs RA (2006) Genetics. No longer de-identified. Science 312(5772):370–371

Mello MM, Wolf LE (2010) The Havasupai Indian Tribe Case – lessons for research involving stored biologic samples. New Engl J Med 363(3):204–207. doi:doi:10.1056/NEJMp1005203

Moore v Regents of the University of California (1990). 793 P 2d 479 (Cal SC 1990)

Nyholt DR, Yu CE, Visscher PM (2009) On Jim Watson's APOE status: genetic information is hard to hide. Eur J Hum Genet 17(2):147–149

O'Neill O (2003) Some limits of informed consent. J Med Ethics 29(1):4–7

Obasogie OK (2009) Playing the gene card? A report on race and human biotechnology. Centre for Genetics and Society, Oakland, CA

Panoyan L, Lee S, Arar R, Abboud HE, Arar N (2008) The informed consent process in genetic family studies. Genomics Soc Policy 4(2):11–20

Parents of children with darker skin lose IVF appeal. (2011) BBC. http://www.bbc.co.uk/news/uk-northern-ireland-13909127. Accessed July 2011

Paynter NP, Chasman DI, Buring JE, Shiffman D, Cook NR, Ridker PM (2009) Cardiovascular disease risk prediction with and without knowledge of genetic variation at chromosome 9p21.3. Ann Int Med 150:65–72

Pelias MK (2006) Human subjects, third parties, and informed consent: a brief historical perspective of developments in the United States. Commun Genet 9:73–77

Personal Genomes Project. http://www.personalgenomes.org/. Accessed July 2011

Risch N (2006) Dissecting racial and ethnic differences. New Engl J Med 354(4):408–411

Rotimi C, Leppart M, Matsuda I, Zeng C, Zhang H, Adebamowo C, Ajayi I, Aniagwu T, Dixon M, Fukushima Y, Macer D, Marshall P, Nkwodimmah C, Peiffer A, Royal C, Eiko S, Zhao H, Wang VO, MCEwan J, HapMap Consortium TI (2007) Community engagement and informed consent in the International HapMap project. Commun Genet 10:186–198. doi:doi: 10.1159/000101761

Seguin B, Hardy B-J, Singer PA, Daar A (2008) Genomic medicine and developing countries: creating a room of their own. Nat Rev Genet 9:487–493

Taylor P (2008) When consent gets in the way. Nature 456(6):32

Walloo K (2006) Stigma, race and disease in 20th century America. Lancet 367:531–533

Weinfurt KP, Hall MA, Friedman JY, Hardy NC, Fortune-Greeley AK, Lawlor JS, Allsbrook JS, Lin L, Schulman KA, Sugarman J (2008) Effects of disclosing financial interests on participation in medical research: a randomised vignetter trial. Am Heart J 156(4):689–697

Wellcome Trust Case Control Consortium. http://www.wtccc.org.uk/. Accessed July 2011

Chapter 6
Autonomy and Its Limits: The View from Genomics

Over himself, over his own body and mind, the individual is sovereign.
John Stuart Mill, *On Liberty* (1864).

Abstract The important notion of autonomy has already been introduced. In this chapter, we look more closely at different philosophical and everyday understandings of the concept. This includes contrasting autonomy used as a descriptive term and as a normative term; negative and positive approaches to autonomy; the scope of autonomy; and the relation of autonomy to community and other values as well as alternatives to protecting and valuing individuals. Applications of these ideas to particular genomics research projects are illustrated. Important limits to our autonomy are examined through a discussion of the findings of research by Stanley Milgram on obedience to authority which draws out the particular implications of this for research ethics, especially within an institutional context.

6.1 Introduction

We have already seen the highly visible place of the concept of autonomy, utilised as a device to value and protect the individual, in discussions of particular issues arising in genomics research ethics. Now is the time to look further at this important, pervasive, but slippery and controversial notion.

Any survey of the philosophical literature written on autonomy would quickly reveal how much ink has been expended in dissecting and discussing the notion, in relation to both personal morality and political philosophy. Likewise, much work in individual and social psychology also addresses questions concerning autonomy and one aim of this chapter is to bring these two together.

P. Boddington, *Ethical Challenges in Genomics Research*,
DOI 10.1007/978-3-642-23699-0_6,

One initial question is indeed, what exactly is meant by autonomy. As we have seen, it may be assumed that the value of autonomy present in clinical medicine and in clinical research ethics can simply be transposed to the situation of genomics research. However, the extrapolation from one context to another will depend vitally upon how exactly the notion of autonomy is understood, as well as on how it is balanced with other values.

The notion of autonomy has been widely interrogated, and its place in medical ethics more widely has been questioned (Walker 2008), but it is of particular interest to us that genetics and genomic science directly present challenges, both to some of the ways we have been understood as autonomous and separate individuals and to the reliance on autonomy as a normative concept upon which to ground our ethics. We will also examine some deep-seated difficulties with the notion of autonomy via a close consideration of work in psychology that can indeed be seen as an empirical interrogation of the autonomy of the individual. The ways in which discoveries in biological science, in social science, and developments in technology themselves affect our ideas of our own autonomy is part of the complex mix needed to address this topic.

Different ways of understanding autonomy and related terms

Absence from external causation
Actions in accordance with the will alone where it is a necessary condition for moral agency (an interpretation deriving from the ethics of Immanuel Kant)
Actions which are truly one's own
Authenticity
Capacity to know what morality requires of us (a Kantian interpretation)
Dignity
Free choice
Home rule
Integrity
Individuality
Independence
Knowledge of one's own interests
Liberty to follow one's own will
Non-interference
Self-assertion
Self-determination
Self-government
Self-knowledge
Self-mastery
Self-sufficiency
Self-rule
Sovereignty
Responsibility

Sources include *Roget's Thesaurus* (1936), the *Oxford English Dictionary*.

6.2 What Is Autonomy? Philosophical Accounts of the Notion of Autonomy

There has been an immense amount of philosophical work that has examined the notion of autonomy (Christman 2011). What follows is a brief overview of a few of the central issues of most concern to our investigations.

6.2.1 Descriptive and Normative Accounts of Autonomy

It is vital to realise that autonomy is sometimes used as a descriptive term, and sometimes as a normative (value) term. Using it descriptively, to possess autonomy is to have the capacity to exercise free choice; for the individual to be the source of his or her own behaviour. This may involve the ability to gather, understand, and weigh up relevant information, and to arrive at a decision based on this without being under any undue outside influence. Sometimes this is described as being able to act in accordance with one's own authentic wishes. Certain individuals may have their autonomy compromised, and lack, or possess only to certain degrees, the wherewithal to think and act with autonomy. So, for example, babies, younger children, those suffering from advanced dementia, will suffer from reduced or absent autonomy. People under duress, for example, acting under conditions of threat, or in certain states, such as intoxication, will likewise have reduced autonomy whilst those circumstances endure.

Now, it is important that there are different ways in which we might have our autonomy compromised, and some of these cases are clearer than others. It is unarguable that a baby has not developed autonomy beyond being the instigator of crying, smiling, and so on; it is a moot point when an older child develops the capacities that grant it the status of a fully fledged autonomous adult. Some of the ways in which autonomy may be compromised can be explained by reference to the individual: they have developed severe dementia, for example. Other ways refer to social settings or external circumstances which compromised autonomy: lack of information, or clear coercion, count amongst these.

It is also now well understood that in the medical context in particular, people's capacity to act as autonomous agents may be compromised by the setting of authority and dependency that characterises the practice of medicine. Indeed much work has looked at these problems with autonomy both at an individual level, for example, considering how to determine when a person's capacity is reduced, and also at the general social setting of medicine and the operation of power within this. This is greatly important, for when we consider autonomy as a value, it will be a mistake – especially severe for those who place great emphasis on the value of autonomy – to overlook ways in which a person's capacity for autonomous choice might in practice be compromised. We will look more closely at this idea presently.

6.2.2 Mere Choice: Or Reasoned Choice?

On one understanding, autonomy is fundamentally about choice; about who has control. On such an understanding, what matters is simply that the person chooses, without undue or outside influence. The content of the choice is entirely subjectively chosen – also, it must be noted, without much if any analysis of where that choice comes from other than that its origin is in the agent him – or herself.

On an alternative approach, an autonomous decision or action is that which a rational agent would choose. Certain standards of decision making apply. To illustrate this with a currently debated topic: broad consent (Lunshof et al. 2008; Hofmann 2009; Mascalzoni et al. 2008). Some argue that if autonomy is seen to imply that any choices are sound and based upon a full appraisal of the situation, then it has been argued that consent cannot be valid – cannot be genuine, fully informed consent – if it is simply broad consent, where we cannot know the full details of all future research. Rather, full detailed information must be provided, for it to be valid consent, and for it to rest upon a proper notion of autonomy.

Immediately we see how a practical debate may rest upon different philosophical conceptions of the nature of autonomy.

6.2.3 Negative and Positive Approaches to Autonomy

A normative use of the notion of autonomy sees it as something of value to be promoted. Normatively, autonomy may be understood as a negative value: we should not interfere with someone else's autonomous choices, should not coerce, overly influence, or give false information. It may also be seen as a positive value, that more strongly, we should act to enhance, support, and assist someone else's autonomous choices and actions.

This distinction also has great practical implications. Should we actively assist someone considering taking part in research to think through the issues carefully? Or should we simply give them the information and leave the rest up to them? Sometimes issues are even weightier than this. It is often argued that on occasion, preventing a person from taking certain choices that might curtail their future decision-making powers is acting positively to maintain their future, long-term autonomy. For example, this thought may lie behind the standard practice for providing counselling for various serious genetic tests, such as for Huntington's disease, that there must be a time period between requesting testing and the provision of counselling, and the actual test. Since the test results are final, it is considered that a safeguard to people making a decision too hastily is to force them to have a period wherein they can reflect on the best choice. Their "best" autonomous choice may take time. On a negative picture of autonomy, a policy might more simply allow people to choose testing without any hindrance. Yet again we can see the large practical implications of any theoretical stance.

6.2.4 The Domain of Autonomy: Self, Body, Life, Information

If autonomy is understood as some account of control or mastery over the self, then this raises a number of questions (Buss 2008). First of all, what is the self? This may sound like an abstract philosophical question. But the answer to it, and to questions of why we value control of things viewed as personal or as belonging to the self, will have a large impact upon how we think about the ethics of genomics research.

One straightforward answer to what the self is loosely identifies it with the body. The idea that we should have control over our own bodies is widely endorsed and central to ideas of individual liberty. But even this is not straightforward. It raises questions such as: does our right of control over our own bodies extend after death? Does it extend to excised parts of our body? We can think here of tissue samples, amputated limbs, a baby's umbilical cord, and since we are constantly losing bits of our bodies, also hairs, skin cells, and so on.

These questions, some of which may seem trivial to many of us, are very important to others, and can make a large difference to policy. For an example of this, see the UK's Human Tissue Act 2004, or other similar pieces of legislation, where debates have been had over the status of literally microscopic parts of human bodies (Price 2005). There have literally been debates over the moral status of urine. Not all of these are in terms of autonomous control – a concept also frequently drawn upon is that of respect for the body (or for the person). But what is clear is that these debates have a large impact upon the conduct of research.

Take a blood sample, for instance. It is easy to see how a right of autonomous control over one's body can mean that such a sample can be taken only with consent. But once this consent has been given, what, if anything, grounds any rights of control over that sample? This is not a straightforward question to answer – unless perhaps one gives the simple answer that as it is no longer part of the individual's body, they no longer have any rights of autonomous control over it. It is also actually surprisingly difficult to give a strong explanation for why an individual's autonomy gives any grounds for control over the disposal of their corpse (Boddington 1998). These issues are crucial for genomics research and demonstrate amply how it differs from much clinical research. It will be more complex still to ground an individual's right of control over DNA extracted from a blood sample or cheek swab through a process of extraction and amplification through PCR (polymerase chain reaction) in rights of bodily integrity.

Another answer to the question of the domain of autonomous control is an individual life – our plans to live life as we wish, in a way which cannot be simply captured by our control over bodily integrity but encompasses more than this. This then may ground our control over the use of excised samples and DNA. But unless it can be shown that the uses of these will have an impact on an individual's life, this is also hard to sustain. Plenty of people have given samples for research with no impact at all on their subsequent life course. A more sophisticated answer still may be that our DNA is a central part of our identity (see Sect. 2.2.4) giving individuals a central claim over what happens to it. This however as will become apparent is also a very difficult idea.

Another attempt at an answer is that individuals should have the right to autonomous control over information about them. This grounds much of the concern over genomics research which deals in large part with information – as I tried to demonstrate in looking at the major differences between genomics research and clinical research. But we now get into complex territory. The boundaries of a person's body are rather clear – with exceptions of excised parts, and unusual cases such as conjoined twins, and the common case of pregnancy (and we are all familiar with the difficulties of debates about the relative claims of mother and foetus). It is also pretty clear what might ground an individual's right to control their own body. But what are the boundaries of information about a person, and what grounds their rights of control over this information? These questions are much harder to answer and yet they are absolutely critical questions for genomics research ethics. A simple alternative answer to this is that we have rights of control over this information not grounded on autonomy but only insofar as this is necessary to protect individuals from harm. This idea will be taken up in more detail in Sect. 6.2.6.

It is easy to see immediately the range of problems. A great deal of information is inescapably in the public domain – as soon as you step on the street, others can tell approximately how tall you are, how much you weigh, how old you are, etc. Much more information can be found on many individuals by doing simple Google searches – often a vast amount, easily discoverable especially for those with slightly unusual names or circumstances. Many corporations who collect vast amounts of commercial and consumer information know things about me even I do not know. Avoiding such information leak is really hard work. Much information is rightly in the province of the government – taxable earnings, date of birth, etc. Much information is rightly in the province of employers – your qualifications, any relevant recordable criminal offences, etc. So, what information about us do we have any right to control – and what information can we control anyway?

One answer to this which many would like to give is that we should be able to control any information about us that is personal, or particularly private, and one key candidate for being such information is anything that gets to the core of our identity – bingo – genomics information. (An additional and separate way of grounding the importance to the individual of genomics information is its relevance to health, surely of great importance to both our bodily integrity and also to our life plans.) But this immediately then raises a central and inescapable problem in genomics. We share nearly all our DNA information with other human beings. (Actually we share quite a lot of it with nematode worms.) We only differ from each other in about one in every thousand base pairs.

But what about these points of variation? Each individual who is not a monozygotic twin has a unique combination of DNA variation. However, it is critical to note that even this overlaps with the unique patterns of information held by others. DNA is a code – packaged in readable, organised ways. The depth of this problem is great. The information about an individual may not just be relevant to another. It may quite simply *be* information about another. An individual's mitochondrial DNA, excluding occasional mutation, *just is* information about the mitochondrial DNA of all individuals in the same maternal line of descent. Likewise for Y chromosome DNA which follows paternal lines of descent from father to son. If I want to investigate certain aspects of my paternal lineage, the way to do it is to get my brother David – or my brother's son Sam or my cousin David who is the son of my paternal uncle – to have his Y chromosome analysed. That information derived from my brother or my nephew or my cousin *just is* information about my ancestry:

and about other family members too. Furthermore, probabilistic information about others can be gathered from an individual's DNA – and this can be used to, for example, track individuals "of interest to the police" forensically.

This is a central challenge of genomics and genetics. We think in delving into our genome we are excavating close to the foundations of our own, unique individual being. This seems to be the very territory of autonomy. But what do we find? That we are essentially embodied beings, born of woman, born of flesh that comes of flesh that comes of flesh. To found our ethics then too clumsily on the fiction that we are essentially disconnected individuals suddenly seems crazy.

6.2.5 Autonomy, Community, and Other Values

Autonomy is not the only value. Virtually everyone values autonomy, at least to some extent, unless one is a supreme totalitarian. But it needs to be balanced with other values. The value of individual autonomy can conflict with more social values: the autonomous actions and decisions of individuals can lead to conflict with those of others. The balance between the individualised value of autonomy and more social values is drawn in different ways by different thinkers, and it is fair to say too that it is broadly characteristic of different cultures to find the balance at different points. We saw earlier, for instance, how there is an issue about how to balance the interests of the individual and of the community in genomics research.

Harms, Benefits, and the Greatest Good: Utilitarianism and the Individual

Utilitarianism concerns itself with producing benefits and minimising harms, but crucially these are measured across the whole population. Hence, there are grave potential concerns for individuals on any approach to ethics which simply aims for the values of beneficence and non-maleficence. It is easy to see why: there are many occasions on which the interests of a greater number of people may be maximised by ignoring the interests of an individual or a minority group. There are many occasions in research where one could very persuasively argue that going against the strict letter of informed consent for research on anonymised samples would in fact pose no harm to the individuals; there are other occasions when one could argue that indeed, any such minimal harm was justified by the benefits of saving resources and the knowledge gained.

It is common to think that there may well be justification for acting in ways that may not be best for certain individuals for the sake of the group, especially if these are minor or hard to avoid. However, utilitarianism may have no defence against expecting large sacrifices from individuals for the greater good.

(continued)

These sorts of considerations are a primary reason for introducing a means to protect the individual and to place limits on the means we can use to achieve a greater overall benefit. Notions of respect for persons are one such device (Crisp 1997; Smart and Williams 1973).

Individual and Community Consent in Genomics Research Projects: Some Innovative Practices and Ethical Strategies

There are many examples of research groups which are taking steps to incorporate improved practice in many areas of ethics which prominently include attention to the quality and nature of informed consent. Projects which pay attention to the notion of community consent and consultation can be seen to be trying to balance community and social values with individual autonomy. This is not always an easy balance to strike but there are many examples of impressive attempts to grapple with this issue.

Innovative ethics and communication with individual recruits: The Personal Genomes Project aims to enrol 100,000 individuals willing to have their genome sequenced and made publically available. This therefore departs from standard promises of confidentiality in medical research. Enrolment in the project requires passing a test of understanding of genomics and its implications, and of the implications of taking part. The project includes an evaluation of the experiences of those participating. It is currently based in the U.S. but potentially expanding internationally. It is one example of an increasing number of research projects developing accessible and informative websites with information not just on the science and its application but on ethical issues.

Prior community consultation: The International HapMap Consortium aims to produce a haplotype map of the human genome which shows patterns of common human DNA sequence variation, to be made freely available to researchers as a resource to aid in further genomics research.

Samples are made freely available to researchers, without identifying information but labelled with population of origin.

Potential ethical problems of this strategy identified by the HapMap consortium include the possibility of stigma (see text box p. 98) and the possibility that division into different population groups, necessary for methodological reasons connected with statistical differences in patterns of genetic variation, might be used by some to reify the notion of distinct human races.

An extensive ethics consultation with communities in the populations involved was carried out to ascertain community concerns prior to the start of the HapMap project. Issues covered included local sensitivities to questions of ancestry and the naming of population groups (HapMap Consortium 2004).

Dedicated ethics team: MalariaGEN aims to discover natural mechanisms of resistance to malaria through genomics research. The project has its own ethics team to address ethical issues that arise from such collaborative global work in the genomic epidemiology of malaria. This team aims to identify, analyse, and respond appropriately to ethical issues arising in sample collection, and to share good practice more widely throughout the research community. Some problems are tackled on an individual basis and in some instances policy is formed, drawing upon empirical research aimed at illuminating best practice. Policies evolved include those on data release, data sharing, consent, and re-linking data (de Vries et al. 2011).

6.2.6 *Alternative Ways of Valuing and Protecting the Individual Not Based on Autonomy*

Autonomy may be valued as a way of grounding the value we find in individuals; as a way of offering the individual ethical protection. But there are alternative ways of doing this. If you wish to value the individual, autonomy can be contrasted to notions of beneficence and non-maleficence. Recall our discussion of the importance of informed consent. This may be grounded on the value of autonomy, or alternatively on the claim that individuals are best placed to judge whether they will benefit or be harmed. We saw just now the difficulties of trying to explain why a person should have control over personal information using the notion of autonomy. A simpler approach might be to refer to notions of harm and benefit. If information may be used to ends that detract from a person's well-being, that cause them harm, then that person should have a right of control over that information. This could have practical implications, in that if the individual could be protected from any harm, they do not have such extensive rights of informational control. However, attention solely to harms and benefits over a population can have very deleterious effects for individuals.

If consent is understood as primarily a device for ensuring the protection of best interests and the avoidance of harm, then different views of broad consent are possible too. There might be a concern that without adequate knowledge of future uses, an individual may not be in the best position to judge his or her own best interests. However, some have suggested supplementing broad consent arrangements with governance structures such as ethics committees with participant representatives as a mechanism to shore up the ethical protection of informed consent and hence to fulfil the primary ethical goal of avoiding harm and protecting best interests (Kaye and Stranger 2009). If the primary purpose is autonomy, and if autonomy is conceived as requiring all the relevant information, this supplement to consent would undermine its purpose, rather than support it.

Complications abound, however. One might consider autonomy is saved by consent to governance if this is equivalent to autonomy by proxy. Compare the situation in representational democracies, where voters exercise a kind of political autonomy by electing representatives to government. In reply, it might fairly be said however that this does represent a quite different notion of autonomous control than any direct autonomy.

Balancing Individuals and Society: The Difference Principle
A major question in ethics is how we balance the moral claims of the individual and those of society. There are a multitude of occasions on which these may clash. The value attached to individual autonomy can be seen to be one prime way in which the claims of the community are constrained, to ensure that individuals are not forced to pay too high a price for the sake of others, and that there is a domain of individual privacy within which the individual can operate freely.

Additional or alternative ways of protecting individuals have been devised. Many focus on those who most stand to benefit from protection, the worst off in any given society. John Rawls, in his influential book *A Theory of Justice*, expounded this in his difference principle (1972). When social policy is implemented, there are often "winners" and "losers". The difference principle states the "intuitive idea that the social order is not to establish and secure the more attractive prospects of the better off unless doing so is to the advantage of the less fortunate" (p. 75).

There have been many variations and additions made to this principle but some version of it is extremely influential in public policy as well as in popular debate. A standard question of critique of any governmental intervention is precisely how it will affect the worst off in society. This is an expression of the enduring moral concern with the most vulnerable.

6.3 Autonomy in Research Ethics: Research Upon Autonomy

I will now examine a group of experiments of iconic relevance to our subject of inquiry, for they are indeed experiments on ethics, and in particular, on autonomy, moral choice, and moral authority. I refer to Stanley Milgram's studies of obedience, described and discussed by him in his book, *Obedience to Authority* (1974). What these experiments have to teach us bears much repetition; moral philosophers, as well as those engaged in the practice of research, would do well to pay them great heed. Many of the lessons of these experiments have been demonstrated and corroborated in the work of others. Nonetheless, his experiments perhaps uniquely draw together a plethora of salutary lessons. There are multiple lessons to be learnt, and this study and others like it can be drawn on not just to raise problems but also begin to suggest possible solutions. For now, I discuss issues of relevance to autonomy (and in particular to the use of informed consent as a device to protect research participants). We will later look at the wider implications for ethical regulations within an institutional context. These are concerns of the gravest importance for our investigation into research ethics.

6.3.1 Milgram's Obedience Experiments

As is well known, recruits were told that they were going to take part in a study of the effects of punishment on memory. Subjects were ordinary members of the public, aged 20–50 years, who answered a newspaper advertisement to take part in the study for $4.00 an hour. The recruits thought that they were randomly assigned to the position of "teacher", but in fact the person assigned to be "learner" was a trained stooge (for many of the studies, a "mild mannered, likeable" middle-aged man). The "learner" was instructed to learn word pairs, and for error, the "teacher" had to administer electric shocks, beginning at a low level of 15 V (which feels just like a mild tingle) increasing by 15 V increments up to a maximum of 450 V, announcing the voltage level before each shock. The dial on the apparatus showed both the voltage and descriptions of level of the shock each of which covered four 15-V levels, from "slight shock", "moderate shock", through to "intense shock", "extreme intensity shock", "danger severe shock", and for the last two levels (435 and 450 V) no words but simply "XXX".

The "learner" did not in fact receive any shocks at all, but was acting. As the shock level increased, he cried out as if in pain, eventually asking for the teacher to stop, and in some versions of the trials, he called out that he thought he was having a heart attack, following which he fell into silence. If the "teacher" objected to the experimenter in any way about the shocks, the experimenter had a set list of encouraging remarks to make, progressing in turn as needed, from "Please continue", "The experiment requires that you continue", "It is absolutely essential that you continue" to "You have no other choice, you must go on". If a specific question was asked about injury to the "learner" the response was "Although the shocks may be painful, there is no permanent tissue damage, so please go on".

Many different variations of this experiment were carried out, altering factors such as the physical presence of the experimenter, the appearance of the learner, the presence of another "teacher" who complied or not with the experimenter's demands. This has also been repeated throughout the world, on many occasions. In trial after trial, the same broad result has been found. Although some objected and refused to continue to shock the "learner", the majority of subjects, often the overwhelming majority, continued to administer "shocks" to the "learners" up to the maximum level – 450 V, a massive 210 V higher than the domestic electricity supply in many parts of the world.

Philosophers throughout the ages have grappled with the question of what can be expected morally of people, and what is the least we can ask of each other. David Hume, for instance, famously asked, "Would any man, who is walking alone, tread as willingly on another's gouty toes, whom he has no quarrel with, as on the hard flint and pavement?" (1957). Many have likewise claimed that a minimum requirement of morality is that we do not inflict pain on a stranger "with whom we have no quarrel". Milgram showed us, however, that many of us would do precisely that.

6.3.2 Moral Conceit: What We Do Not Know About Ourselves

I have discussed earlier the dangers of supposing that "we" would never do that: it can occlude the chance to learn from the mistakes of others. Many ask how this experiment ever got ethics approval. Much of the comment that these experiments now attract consists largely in high-minded complaints that they were ever conducted, and of claims that "of course" one could never conduct such experiments now. The irony of such claims is underscored by one of the chief findings: the endorsement of many centuries of wisdom that has concluded how little we know ourselves and how little we know of how we will act when put to the test.

Prior to the experiment, Milgram asked a wide range of people, including professionals such as psychiatrists, for their predictions of what they thought people would do. Repeatedly, the answer came that a very few aberrant individuals might progress to danger level, but that most would refuse to obey beyond the lower levels of shock. Wrong. Repeatedly, in different permutations of the experiment, a large proportion of subjects continued, in response to the promptings of an experimenter in a grey lab coat, right up to the highest level.

This then provides one easy explanation of why the experiment obtained ethics approval: quite simply, nobody predicted the results. Nobody thought that ordinary people recruited through a newspaper advertisement would carry this out. Nobody thought that the experiment would produce subject after subject after subject who had to face up to the fact that they had willingly taken actions that they (falsely) believed to result in pain and injury, or even death, of another human being.

The experiment was explicitly carried out in the aftermath of World War II. Milgram, especially as a Jew, was concerned to study what conditions could possibly give rise to the horrors of Nazi Germany. "Oh, *we* wouldn't do that, it's something about *them*!" is an obvious response to such extreme events. Unfortunately, it is an erroneous one. Indeed, it is actually only now that we have the results of Milgram's experiments, and other seminal studies such as the Stanford Prison Experiment that many pronounce with confidence that such studies should not take place. This is a lesson to take away: that our understandings of what is ethical change in response to experience. It is only as a result of such research that we have gained the insights into people's behaviour that lead us to expect such an outcome: so, prior to such an experiment, how could we reasonably anticipate such dilemmas for the subjects? This is a general point: that repeatedly, advances in knowledge lead us to reappraise and reconsider ideas of what is and what is not ethical.

This is a major lesson to be learnt, but one which humanity should surely have learnt by now. We suffer from lack of moral self-awareness. We act in ways we do not predict. We act in ways that go against what we, on reflection, believe we should do. It is necessary to think about this in relation both to the position of the research subject and the researcher, and to think about what it means for the notion of autonomy.

6.3.3 The Fragility of Autonomy

As we have also seen, an original goal of research ethics is to protect individuals from the misplaced use of power and authority by those conducting experiments, one way in which this is achieved is through the requirement of free and informed consent, and a dominant way in which the underlying values of research ethics have come to be articulated is through the principle of valuing individual autonomy. (The sort of social and psychological research which Milgram and others conduct is recognised as necessitating in some cases that there is an element of deception, to allow the experimental conditions to be fulfilled. In such instances, there must be adequate debriefing after the experiment closes.)

But the Milgram experiments show us just how fragile our "autonomy" is, and how vulnerable it is to the claims of authority of others. Subjects in the experiment obeyed the commands of a person in a grey lab coat whom they had only just met. The chilling ease with which a person can don the garb of authority is a lesson indeed. Milgram had wondered if this was affected by the prestigious institutional affiliation of the experiments, conducted in the highly respected Yale University, so he transferred some of the experiments to an anonymous and rather unimpressive room in a dingy office block in Bridgeport. The results were scarcely affected. The ideology "this is for research", the mantle of a simple grey coat, were sufficient. We must consider the implications this has for building the foundation of the house of research ethics upon the rock of autonomy as the rock seemingly morphs into shifting sands.

Milgram himself took his experiments to undermine the simple view that "the individual is pre-eminently the source of his own behaviour". The experimenter told subjects that he would accept responsibility for the outcomes, if for instance the "learner" was in fact having a heart attack. Milgram argues that in such instances, the subjects were acting not in accordance with their autonomy but in what he calls "agentic mode". They handed over their moral autonomy to the experimenter who thus took the role of moral authority for them. Repeatedly, subjects to the experiment justified their continuation on the grounds that they were not responsible for the set-up, the experimenter was.

Those who were most likely to continue the experiment up to high levels of shock were more likely to attribute a greater degree of responsibility to the experimenter than to themselves, as contrasted with those who finished earlier, who tended to see themselves as possessing a greater degree of responsibility for what occurred. According to Milgram's analysis of the experiments, subjects who continued to shock were not actually acting in accordance with autonomy – the supposed cornerstone of research ethics – at all. The authority of the experimenter and of the institution, which the protocols of research ethics have been designed to limit, surfaces above the layers of regulation, the international protocol, the institutional review boards, the consent forms, the bureaucracy, the dogma of personal liberty, like Venus rising resplendent from the ocean, garbed in a grey lab coat and quietly repeating the mantra of scientific progress: "the experiment must continue".

6.3.4 Consent, Authority, and Institutions

It is also fascinating to read the detailed accounts Milgram gives of some of the subjects' accounts following the experiments. One subject was a Professor of Old Testament, who refused to shock after 150 V. Intriguingly, his account was not of disobeying authority but of switching from one authority to another: that of the "learner". He stated in reply to the experimenter's claim that there would be no permanent tissue damage, "Well, that's your opinion. If he doesn't want to continue, I'm taking orders from him" (p. 49). In discussion during the debriefing, he commented, "If one had as one's ultimate authority God, then it trivialises human authority" (p. 50).

The experiment itself gives us plenty of reason to be wary of making the notion of autonomy, and the notion of informed consent, do too much ethical work. Predictions of our own behaviour are consistently wrong, as shown by the predictions of the sample of experts Milgram consulted before his experiments began. Predictions that we shall act with autonomy are often mistaken. Many of the subjects themselves appeared to relinquish their own autonomy.Whither then, autonomy?

One thing that it is vital to note is this: subjects may not have consented to the real agenda of the research, *but they did consent to deliver electric shocks to a complete stranger*. This in itself is a startling finding.

Especially in the context of research which almost always takes place within a prestigious institution such as a university or a hospital, the powers of persuasion and the implicit suggestion that taking part in research is a "good thing" are omnipresent. This at the same time gives extremely good reason for the safeguarding of research subjects by the adoption of informed consent policies, and the formal signing of informed consent. By the same token, however, it also should make us wary of the operation of the rigmarole of informed consent, and wary of how easily it is hijacked, whether consciously or not, and how easily it may be degraded, how easily the autonomy on which it is founded is handed over to another authority, to the institution or to the experimenter him or herself.

Informed consent is only one aspect of research ethics. One conclusion from our consideration of Milgram's work is that we should take perhaps even greater care to ensure we are not placing undue weight on it, nor that we are placing undue weight on the capacity of individuals to act with autonomy. Use of informed consent has to be implemented and supplemented by a careful awareness of how easily people will do what is required of them, and how easily they will do this even if it goes against what they believe. The theoretical dogma that the ground of informed consent is the autonomy of the subject also has to be viewed with caution.

Milgram examined "binding factors" that kept subjects obeying the experimenter. These include the sequential nature of the actions and the anxiety of going against the social obligations of the situation. The power of socialisation can be illustrated by drawing our attention once more to what was in the balance: as Milgram expounds in his extensive analysis, anxiety about being impolite by failing to complete a task undertaken in compact with a stranger in exchange for four dollars was weighed against subjecting another person to the risk of death; and usually won. These factors of socialisation are omnipresent. They at once give reason to protect the autonomy of research subjects (and others) and give reason to consider that the exercise of this autonomy will be limited. This then means that in order fully to respect the autonomy of individuals, we need to look fully at sources that might act against it.

Against this, we must also consider that trust in others, including moral trust, is often justified, or at least, has some claims to merit. Merited trust in individual researchers, in institutions, and in the whole practice and conduct of research is a vital aspect of research ethics. The operation of an institutionalised framework of research ethics can, and should, help to maintain a legitimate trust. This topic in itself is worthy of extended consideration: both in uncovering persuasive forces operating on our beliefs and actions and in examining their legitimacy or otherwise. But for now it shall be observed that such merited trust – such legitimate sources of moral authority – cannot be said to interfere with the autonomy of individuals in the same way as illegitimate or unmerited sources of authority.

6.3.5 Consent, Conscience, and Persuasion

We also can take from the Milgram experiment another lesson, for in fact it provides a potent clue about how to go about getting someone to do something they would never initially agree to. If you asked a group of people to deliver 450 V shock to a stranger, it seems highly unlikely that you could get them to agree. (I don't think Milgram actually carried this out, but I have asked seminar audiences if they would do this. They all say "no".) So how could you do it? A little bit at a time, in 15-V steps. Milgram has elegantly demonstrated to us the childish simplicity of the building blocks of evil. The ease of obtaining compliance should also be food for thought for any who blithely rely on the signing of a consent form to provide ethical clearance for research.

Interestingly, research ethics committees often attempt to protect the interests of research subjects by careful circumscribing of consent conditions that limit it to a particular narrowly described programme of research. Obtaining consent to the use of samples and data for subsequent projects often then proceeds in a piecemeal manner, bit by bit. Protocols for long-term projects in fact now often consider that this method provides superior protection (Mascalzoni et al. 2008). *Yet Milgram's work suggests that this is precisely how to obtain consent to something that the subjects would not have agreed to, had they been told at the outset what the later requests would involve*. This underlines the need for safeguarding research subjects from harm as well as operating a policy of informed consent. It also suggests that a one off consent may in at least some instances be more valid than a step-wise consent procedure.

Boiling Frogs

Apologies to animal lovers: this unpleasantly named persuasive technique refers to the supposed phenomenon whereby a frog placed in a pan of hot water will of course jump out, but if placed in cold water which is very gradually heated, will remain in place until cooked. It is debated amongst scientists whether this is in fact the case (and please do not try this at home). But the metaphor has been used in a variety of contexts. One major use is to indicate a method of getting people used to a change of practice, by introducing changes imperceptibly so that the large change from one state to another is not perceived or resisted. In terms of the introduction and impact of new technology the relevance of such a metaphor is obvious, and those attempting to point out the effects of technological advances upon our lives and on our values are often in effect making such a point.

The metaphor of course relies on the undesirability of the end point of being boiled alive; other metaphors that follow broadly similar courses, such as weaning a baby from milk to adult food, do not of course carry such a connotation.

A salutary thought for considering the role of informed consent in research is this: many of the subjects, asking why they had continued to administer shocks to the learner, cited his agreement to take part – "he consented, didn't he? He's got to see it through to the end". This is an argument in effect to take a subject's consent as so robust that it overrides over sources of morally relevant consideration. Milgram's experiments are worth our careful study for this point alone: that it should give us food for thought over whether consent to research participation is in itself sufficient to render such research ethical. How much authority does the fact that a person consented really give us?

Work such as this shows us that the urge to obey authorities and to follow orders is strong and easily triggered. Especially within institutional settings, and with the trappings of authority that belong to science and to medicine, this has profound implications for informed consent to research. The very ease with which people can be persuaded to act provides a sound and pressing reason for insisting on informed consent to participation in research, in order to counter this easily won obedience.

The inertia of conscience and of autonomous action is also demonstrated by Milgram. It was extremely common for subjects to become rather agitated and upset during the experiment. In interviews afterwards, subjects generally reported that they experienced great stress at delivering apparently painful shocks to the "learner", when the experiment required that they continue. The crucial point for us is this: that even in those who abandoned the experiment, there was consistently a significant delay before the continued promptings of a subject's conscience won out. Conscientious objections struggled to operate against the protocol of the experiment which formed an easily acquired mask of authority. There was a large gap between ethical beliefs and action. Milgram used his experiments to question the assumption that "the individual is pre-eminently the source of his own behaviour".

It is important also to recall that so long as you do not think there is anything wrong with inflicting a mild tingle on another person, who seemingly agreed to this, a genuine question does arise: when to stop? At what point is it entirely clear what a subject in Milgram's experiment "ought" to do? Where exactly would you draw the line at the maximum amount of shock that anyone "ought" or "ought not" to deliver?

Indeed, Milgram did not consider that the subjects' moral sense had simply *disappeared*: it rather *shifted course*, from following through a basic moral tenet that one should not inflict harm on an innocent person, to attempting to live up to the expectations of someone in whom one had invested authority, and to live up to a task one has agreed with that authority to undertake.

6.3.6 The Institutionalisation of Consent

It is fascinating to note that the Milgram experiments grew directly out of the same milieu that spawned much of the ethical regulation of research: the impetus, post-World War II, on the one hand, to *understand* and as a companion to this, to *guard against* the horrors of the Third Reich. A major question was why so many people complied with the regime. Again and again, an explanation for actions was heard: "I was only following orders" (Muller-Hill 1998; Arendt 1994) . How did this happen? Was this something peculiar to German culture at the time, for instance? Surely people in democratic societies where individual freedom is valued would do no such thing? Sadly, the answer is clear. People may well follow orders that conflict with their conscience, calmly considered.

But we are faced with a conundrum: the regulatory machinery that has grown up in response to the abuses of the past is premised upon the notion that people are autonomous individuals who make free and rational choices. Yet the investigative responses of social science research, such as Milgram's, have found precisely the opposite: over and over again, it is found that it is easy to "shut off" the autonomous individual inside us all, to induce people to do what they are told, to act in ways that go against individual conscience, to elicit behaviour that does not stem from the transparent, coherent, open workings of the rational mind.

We are presented with two conflicting images of humanity: as rational, free, autonomous: and as subject to forces beyond our control. We now face the task of keeping both in view at once.

One solution to this conundrum lies in appreciating the difference between fact and value. The requirement of consent to research, premised on the idea that we are rational, free, and autonomous, is premised upon not so much a fact about humans, as a value: we have to regard people as individuals in their own rights, with the right and the power to mould their lives and their destinies. Giving the right of informed consent to research then expresses this core notion of respect to the individuality of humanity: it is a key part of what philosophers have called the "separation of persons" (Downie and Telfer 1969).

Valuing each of us as individuals is vital to preventing great moral abuses, whereby individuals are used merely as the tools of the state or of society. But equally vital to understanding our motivations and our actions is the need to keep in view the essential fact that humans are social animals; we respond to others and to cues about hierarchy and authority. Moreover, a great deal of research demonstrates that we lack full awareness of why we do things, and of why we think what we think. Many of our own motivations are opaque to us. If we allow ourselves to forget how easily we can fail to be truly free, fully rational, and autonomous, we will, ironically, fall short of the aim of providing protection for the individual.

If, for instance, we ignore the effects of authority and hierarchy, then a signature confirming consent to research obtained in no matter what circumstances will be fine. It is recognised, however, that participants can be induced to take part in research, that they can be swayed by the authority of institutions and of high status individuals. In recognising such forces, it is acknowledged that we are not always fully autonomous. In other words, it is recognised – or should be recognised – that truly informed consent is hard to obtain, and that consent alone is not enough.

We have seen that one response to difficulties with informed consent is to make attempts to shore up consent, to make it as informed and as freely given as possible. Another response is to add on extra ethical requirements, such as oversight by committees with participant representatives. Counters to such responses include suggesting that this dilutes the informed consent of individuals, and hence is an affront to the value of individual autonomy.

Individual autonomy is something that we rightly strive for, – balanced against other values – but the idea that we are fully autonomous and entirely free individuals is a myth. It is a convenient myth for many reasons, one of which is that it functions as a device of ethical protection. It is a device, one of many in our ethical armoury, which can function as "stop" button to treating humans as means to ends. Embodied in regulation through notions such as the requirement to obtain informed consent, it functions to remind us to respect persons. It is a device for avoiding dehumanisation of research subjects. Used alone, however, it is open to abuse: just because somebody "consented" to something does not in itself make it ethically sound.

There is another way of expressing our conundrum in terms of what it is to dehumanise a person. We can dehumanise if we treat someone as less capable than they are, as less than a free, separate individual with an identity, a personality, and a life of their own. But we can also dehumanise in another way, when we overburden or fail to recognise shortcomings, and hence fail to provide safeguards and handrails to assist our fellow, frail humanity.

I would strongly argue, in conclusion then, that whilst it is important to ensure that consent to research is robust, to chase the notion of fully informed, and fully free consent, is to chase something of an illusion, and that consent functions as one of a range of means to ensure that research on human subjects does not dehumanise, does not take advantage, does not exploit, does not risk undue harm.

6.4 Conclusions

Autonomy can be understood in different ways, and these will have an impact on how it can be applied ethically and how in different situations there might be different ethical answers.

The value of autonomy has to have limits. What those limits are will depend crucially on how autonomy is understood and articulated conceptually.

We need to understand how the value of autonomy is based upon assumptions about the individual self, and the relation of the individual self with others. These are assumptions that we need to be aware of and which we might wish to question. Indeed, work in genomics itself helps to undermine some popular conceptions of the extent of individual autonomy in highlighting what we share with others.

If we value autonomy, we should take seriously how easily it can be eroded. This involves looking at individual and social psychology, and at the institutional surroundings of our actions.

6.5 Study Questions

1. It is important to think through why and in what contexts you value autonomy, and when you might accept limitations on your autonomy, and for what reasons. In doing this, it is invaluable to think through these questions for a diverse range of examples. For example, consider this issue. In some countries, for example, Australia, it is compulsory to vote. Unless you have a good reason, such as illness, you can be fined a (fairly modest but not insignificant) fine for failing to vote. Is this an infringement on a person's autonomy to decide not to participate in the political system? Or is it a device to ensure that we do in fact act as autonomous agents contributing to that very system? (Note that of course, the ballot is private – one can spoil one's ballot paper or leave it blank, so one is not

actually forced to vote for one of the candidates on offer, only required to go through the formal process of voting.)

2. Our autonomy can be diminished in various ways. Think about the impact of these different situations on an individual's autonomy:

   Enrolment in a national database of DNA for medical research is made compulsory under national law.

   A highly successful advertising agency conducts a persuasive advertising campaign to get people to volunteer for a national biobank.

   A genomics research project offers recruits a financial reward equivalent to 2 week's unemployment benefit for participation.

   A genomics research project offers full feedback of individual results to all recruits.

   A genomics research project needs to recruit family groups, and a close relative who has volunteered nags you to enrol as well.

   Giving census information to the government is compulsory and punishable by a significant fine.

3. There can be few people who have not in some way benefited from the fruits of medical research. Do we therefore have any obligation to contribute to such research? Are those who do not contribute "free riders", or are they simply exercising their autonomous choice? Does it make any difference to how you answer this question if you are thinking about a system of private health care or a system of public health care?
4. A system of community consultation is proposed for a large genomics research project that hopes to enrol certain identifiable population groups. How should such a consultation process proceed? How should such a process respond to differences of opinion within such a group? How should such a consultation process go about identifying appropriate authorities within the group?
5. One of the subjects in the Milgram experiments who refused to continue to shock was a Professor of Old Testament Studies who said "If one had as one's ultimate authority God, then it trivialises human authority". Does following a moral code derived from a religion undermine one's own autonomy? Does following any kind of moral code undermine one's autonomy?

6.6 Coming Up Next

We have seen the prime importance of considering the ethical claims of the individual and also of the community, and the difficulties of articulating how to value the individual through the notion of autonomy, and the difficulties of balancing these claims. We have seen how an individual's capacity to make considered choices and moral judgements may be compromised very easily by social circumstances. We have also previously considered the setting of ethical regulation, and ethical decision making, within a social setting. The location of the individual in society cannot be overlooked, and needs to be emphasised. To this we now turn.

References

Arendt H (1994) Eichmann in Jerusalem: a report on the banality of evil. Penguin, London
Boddington P (1998) Organ donation after death – should I decide, or should my family? J Appl Philos 15(1):69–82
Buss S (2008) Personal autonomy. Stanford Encyclopedia of philosophy. Stanford University, Stanford
Christman J (2011) Autonomy in moral and political philosophy. Stanford Encyclopedia of philosophy. Stanford University, Stanford
Crisp R (1997) Mill on utilitarianism. Routledge Philosophy Guidebooks. Routledge, London
de Vries J, Bull S, Doumbo O, Ibrahim M, Mercereau-Puijalon O, Kwiatkowski D, Parker M (2011) Ethical issues in human genomics research in developing countries. BMC Med Ethics 12(1):5
Downie RS, Telfer E (1969) Respect for persons. George Allen and Unwin Ltd., London
HapMap Consortium (2004) Integrating ethics and science in the International HapMap project. Nat Rev Genet 5:467–475
Hofmann B (2009) Broadening consent and diluting ethics? J Med Ethics 35(2):125–129
Hume D (1957) An inquiry concerning the principles of morals. Bobbs Merril Company Inc., Indianapolis
Kaye J, Stranger M (eds) (2009) Principles and practice of biobank governance. Ashgate, London
Lunshof JE, Chadwick R, Vorhaus DB, Church GM (2008) From genetic privacy to open consent. Nat Rev Genet 9(5):406–411
Mascalzoni D, Hicks A, Pramstaller P, Wjst M (2008) Informed consent in the genomics era. PLoS Med 5(9):e192
Milgram S (1974) Obedience to authority. Harper Collins, New York
Mill JS (1864) On liberty, 3rd edn. Longman Green Roberts Longman and Green, London
Muller-Hill B (1998) Murderous science: elimination by scientific selection of Jews, Gypsies and others in Germany, 1933–1945. Cold Spring Harbour Laboratory Press, New York
Oxford English Dictionary (2011) Oxford University Press
Personal Genomes Project. http://www.personalgenomes.org. Accessed July 2011
Price D (2005) The human tissue act 2004. Modern Law Rev 68(5):798–821
Rawls J (1972) A theory of justice. Oxford University Press, London
Roget PM (1936) Thesaurus of English words and phrases. Longmans, Green and Co., London
Smart JJC, Williams B (1973) Utilitarianism for and against. Cambridge University Press, Cambridge
Stanford Prison Experiment: a simulation study of the psychology of imprisonment carried out at Stanford University. http://www.prisonexp.org/. Accessed July 2011
Walker R (2008) Medical ethics needs a new view of autonomy. J Med Philos 33(6):594–608

Chapter 7
The Social and Institutional Setting of Ethics

Abstract One conclusion so far is that there is a need to examine in more detail the institutional context in which ethics is regulated and in which ethical decisions and rules are made. This chapter discusses the social context of science, including financial issues, scientific teamwork, science and the wider society, and authority within institutional contexts. Lessons are drawn from the well-known example of the Tuskegee Syphilis study. Social influences on moral action are discussed and work on administrative ethics is introduced. With a focus on ethical issues in research, findings from philosophical ethics and from empirical work in social psychology are put together to produce a discussion of ethical hindrances to moral action, together with suggestions about how to address these problems.

7.1 Introduction

We have seen in our discussion of autonomy in the last chapter how careful attention has to be given to the circumstances in which autonomy might be eroded; here we turn to consider more deeply the social and institutional influences on belief and behaviour which may undermine ethics; but which may also be turned to good effect to address ethical problems. This is an area of great importance in applying ethics to practice, and yet which has received little sustained systematic attention within the field of applied ethics. There is however, much work in the social studies of science, and in social and moral psychology, to draw upon in our efforts to improve the ethical regulation and practice of science.

As we have seen, there has been a great deal of attention to ELSI – to the ethical, legal, and social implications of genetics and genomics. Despite this linking of ethics and social issues, questions of ethics are often still presented as stand-alone, abstract principles or codes of conduct. The way in which ethics in a medical or research context may be taught often reinforces this abstraction. It would be the gravest irony if the existence of an institutionalised ethical regulation acted to divorce individuals or research groups from their own moral responsibility by

P. Boddington, *Ethical Challenges in Genomics Research*,
DOI 10.1007/978-3-642-23699-0_7, © Springer-Verlag Berlin Heidelberg 2012

parcelling it off to the system of ethical oversight. At the very least, it is important to make sure that the very existence of rules and regulations in ethics does not act to stifle ethical thinking or the exercise of conscience and moral insight.

7.2 Science in a Social Context

If we approach ethics primarily through regulation, rules, and principles, it is easy for our thinking to become too focused on the abstract. We need to balance this against practical grounded realities, and this includes keeping a watchful eye on the social context. Whilst science aims at an objective account of the empirical physical world, the practice of science is a social institution taking place at particular locations and at a particular time in history.

A great deal of work is done in the social studies of science and technology and here this will only be touched upon. Much of this work concerns itself with examining how the social setting of science impacts upon its operation and in particular its claims to uncover objective facts about the world. Much of it is also highly relevant to the ethical regulation and conduct of science. In his book, *The Ethics of Science*, David Resnik argues that it is clear that the research environment of science contributes to much misconduct, providing many examples of how this can happen (1998). A particular interest is not simply the research and social environment of science, but the existence of the apparatus of research ethics regulation itself. This indeed was a question raised right at the start of this book, when I considered my dilemma of trying to operate as someone working in ethics, yet attached to a particular scientific consortium.

To start with the obvious: science is carried out by particular people in particular contexts. The amateur scientist has been a long and honourable history, and there is still a great deal of involvement by amateurs, for instance, in the monitoring of changes to the weather and the natural world. However, most science, especially the science we are considering, is carried out by professionals; people who work in particular institutions, under particular economic, social, and political circumstances.

7.2.1 Science and Money

Extracts from criteria for appointment, senior level academic positions in biomedical sciences, collected from different advertisements July 2011 from jobs.ac.uk:
"The ability to manage, and generate funding to support, a substantial research team"
"he/she is expected to have a continuing track record of external research grant funding over a number of years"
"You will have ... an internationally competitive publication and grant track record"

Money is an extremely important factor in the conduct of science (Resnik 2007). The source of this money cannot be overlooked in any full consideration of ethics in science. Much funding for science is public funding derived from governments. In the realm of medical research in particular, much funding is charitable, from non-governmental organisations with internal control over their funding decisions, with altruistic and publicly minded ideals. In addition, a great deal of funding is also in private commercial hands; for example, from pharmaceutical and biotechnology companies. Money is always limited; there are always questions of resource allocation; it is of course a powerful motivator, a powerful driver and incentive, and therefore potentially a powerful corruptor. In many cases, he who pays the piper calls the tune. Those with the capacity to fund science can therefore have a great deal of influence over the direction and conduct of scientific research.

However, regimes of accountability do of course exist. Especially where money is from the public purse, there will be some level of scrutiny over how and where it is spent, and general calls for economic efficiency. Constraints also operate on commercial sponsors of scientific research, including the drive to turn a profit and to answer to shareholders, but also to maintain reputation and to be seen to be producing good quality research. There is a direct link between scientific standards and ethics in the translation of science; basing treatment decisions on poorly conducted science, or science which biases one commercial product over another, is of great concern to patients; this may mean they are paying for a product they don't need, paying more for a product no more effective, or receiving ineffective or even harmful treatment (Medawar 1992).

In recent years, there have been a number of vigorous calls on commercial funders to demonstrate practices that maintain scientific standards. An example of this is the registration of clinical trials in order to prevent pharmaceutical companies (and other with an interest linked to gaining particular findings) from selectively publishing only the results of trials providing support for their own products (CONSORT), and journal requirements to publish data upon which results are based (Angell 2005). This, in turn, demonstrates the importance of central features of the practice of science: the need for publication of results, especially in well respected, international peer reviewed journals therefore means that the editors of such journals can exert considerable influence on those producing the science and, in turn, on funders.

The source of funding, whether public or private, has had a considerable impact upon the conduct of science in the realm of genetics in particular. This may have far reaching effects. In the public–private race to sequence the human genome funding played a key role. The drive to prevent the private consortium from patenting sequences of the human genome spurred on public and charitable funders to pour money into the public effort and to publish results daily precisely in order to battle private interests and to defend a norm of science as producing knowledge for all and for the public good, rather than something owned by private groups or individuals for their own profit (Davies 2001). This is a vivid illustration of how economics and ethics are intimately intertwined.

Commercial Partners in Genomics Research: Some Ethical Issues

There has been lively discussion of the possible corrupting influences of money in scientific research, especially within pharmaceutical research. The potential earnings and losses of pharmaceutical companies are enormous and could be affected by decisions about the direction of research and the results obtained. The loyalties of employees of pharmaceutical companies are not simply those of a dispassionate seeker of the truth (nor, indeed, may the loyalties of academic researchers be so disinterested).

The regulation of pharmaceutical companies is especially important, given the great, almost insuperable, difficulty of gaining redress via litigation. It is very difficult to bring successful legal action against pharmaceutical companies, for a number of reasons.

Research partnerships involving pharmaceutical companies or other commercial organisations exist. Steps can be taken to reduce the chances of undue influence by commercial partners. For example, decisions about the particular direction of research may be taken at meetings without company representative present.

Where there are commercial partners in research, it is good practice to include this information on the consent form, especially where there is a possibility that commercial gain may result from the research. It must be made clear to subjects that they will not gain from their participation (if this is indeed the case, which is usually so). Some of course may object to taking part in research for nothing when there are those who stand to gain.

This may then present a difficulty with collaboration with other research projects. Such collaboration is common for various reasons, such as increasing sample size or comparing across populations or disease groups. However, the consent for some projects sometimes explicitly rules out any commercial gain from the research. Where this is so, collaboration without re-consent would be problematic.

Funding of individuals also raises serious ethical issues. We are used to considering the power of money in commercial contexts, but no one is free of it. The career structure in science is notoriously perilous; the university sector is now one of the most casualised labour forces there is, and this is especially so in the research sector, where many work from one short term contract to another throughout their careers. The possible ethical ramifications of these factors should not be overlooked. Much attention has been paid to issues of conflict of interest in funding, for example in taking freebies and payments from pharmaceutical companies, and the influence this might have on research; strangely, *additions* to one's salary are seen to influence conduct, whilst the potential influencing power of the fundamental need to earn one's bread-and-butter through securing a basic salary receives scant comment. Yet this need must, on swift reflection, be even stronger than the urge to accept the free trips to desirable locations which are routinely now acknowledged to have the potential to corrupt. Short term contracts can impose tremendous strains and create many practical and personal difficulties for personnel (see Sect. 7.5.1).

7.2.2 Scientific Teamwork

Science is a cooperative activity: even the lone investigator stands, as Newton said, on the shoulders of giants in building findings upon those of others; and most seek to share their findings with the world. Most science is conducted in teams, with these teams often communicating and cooperating with other teams, often internationally. Many of these teams involve those with different areas of expertise working together; trust, reputation, reciprocity, and fair exchange are all important to the conduct of science and of course are ethical concerns too. It is also of considerable importance to observe that scientific practice, especially teamwork, is usually hierarchical. The rocky career structure of the scientific researcher can only reinforce this hierarchy in increasing the dependency of those in lower positions on the references and good favour of those with more power, for example, in roles within a project, and in opportunities for career advancement given, for example, by conference presentations and authorship credits.

Teamwork also involves a specialisation of tasks. The possibility of fragmentation of responsibility and of value presents itself (see Sect. 9.9). Many have commented on the concentration of the moral imagination on a small technical aspect of one's task as a factor that may blind one to the large implications of what one is doing. But specialisation is not inevitably a moral loss and the very fact of the necessity of collaboration and communication gives rise to opportunities for different perspectives – so long as space is made for these.

Who is "the team"? The inclusion of research subjects in aspects of oversight of research is another way in which relevant perspectives might be included.

7.2.3 Science and Wider Society

Science is an activity that takes place within a larger social context; influences from outside science, and from outside the public institutions of science, can also be of supreme importance. Developments in expectations can have a large influence; for example, customers of commercial direct-to-consumer genetic companies are now receiving detailed results of their genome; it is thought that this might influence the expectations of subjects of publicly funded research, and some have suggested that these expectations are rising. The issues of privacy and confidentiality of genomics information in research has already been mentioned (Sects. 5.2.4 and 5.2.5); these issues are influenced by public attitudes towards privacy, which are shaped and changing rapidly with developments from diverse areas such as the wider media and internet technology (Mayes 2002). Many of these influences are outside the control of individual scientists and of the institutional structures of science, yet, nonetheless, must be considered and demand responses. (In Chap. 9, we discuss the extent of researcher responsibilities further.)

Science itself operates according to certain sets of norms, although these are subject to a certain level of dispute; indeed, this is the very subject of this book. However, the norms of science include more than ethics regulation. Resnik, for example, discusses honesty, carefulness, openness, freedom, credit, education, social responsibility, legality, opportunity, mutual respect, and efficiency as well as respect for subjects (1998).

More formalised external influences on science can come from governmental and non-governmental bodies which offer opinions on the conduct and application of science. Among these are the House of Lords Select Committee on Science and Technology, the Human Genetics Commission, The Nuffield Council on Bioethics, and the Human Fertilisation and Embryology Authority. These bodies may have particular remits: some for example are patient support groups (some of which may be funded privately and sometimes with input from pharmaceutical companies), such as the Genetic Alliance; some are critical of the application of science, for example, Genewatch UK; there are also many academic centres which comment regularly on ethics and genomics research, for example, the Foundation for Genomics and Population Health (PHG), the ESRC Centre for Economic and Social Aspects of Genomics (CESAGen), and the Centre for Genetics and Society.

7.3 Institutional Influences and Authority

In considering the ethical conduct of science, it is essential to consider the authoritative influences of institutions and the operation of societal hierarchies. We saw in the last chapter this power at work, and the problems that this can create, for example, in ascertaining that consent to participate in research is given freely.

A conundrum that faces us is that the very regulation of ethics takes place within a hierarchical institutional context where power and authority of course operate. Hence, as well as thinking about what the rules and regulations of ethics in research should be, we need at once to consider how these are enacted in practice. To illustrate some crucial points, I will examine a classic case from the annals of examples of ethically problematic research: the Tuskegee syphilis study (Curran 1973; Jones 1981). It needs to be stressed that what follows is not by any means an exhaustive examination of this much discussed example; I use it here simply to draw a few lessons concerning the operation of authority.

The Tuskagee syphilis study was a longitudinal study of the progress of syphilis over many years. It was thought that syphilis progressed differently in different ethnic groups, and researchers wished to know how it particularly affected those of African origin. Starting in 1932, over 600 African American males were studied. At recruitment, about half were found to be infected with syphilis, although over the course of the study, especially given how endemic it was to the population, some more would also become infected. Recruits knew they were taking part in some kind of study, but did not know that syphilis was under investigation, and were not given their diagnosis. At the time, treatment for syphilis was pretty ineffective, and it was considered that most of those recruited would have been prevented by poverty from receiving any treatment in any case. Those taking part were given various medical tests, had regular treatment for minor ailments, and their families were given burial stipends after an autopsy.

Significantly during the course of this study effective treatment for syphilis became available in the late 1940s to early 1950s in the form of antibiotics. Not only were those taking part not given treatment, local physicians were given the names of recruits so that they would not treat them; treatment would have spoiled the study aims of observing the unchecked, natural progression of the disease. In 1968, concerns were raised and criticism of the study led to a Public Health Service Committee of the Centre for Disease Control (CDC) recommendation in 1969 that the study continue. However, following this, further criticism led to a journalist taking up the issue, whose publications led to another review, recommending that the study be ended in 1972 (Centre for Disease Control).

Naturally, there has been intensive criticism of this study. Consent to participation was certainly not fully informed. Recruits were not given their test results. They were denied the chance of treatment when it became available. It is a classic example of putting the aims of research over and above the needs of individuals. It does now seem quite astonishing that after the highly effective antibiotic treatment for syphilis became available, it was still considered that there was a sufficiently strong scientific need to study the unchecked course of the disease. But, however appalled we are at the behaviour of others, one thing we must resist doing is to fall into moral complacency.

Consider the point of view of those engaged in research. Imagine you have a dispassionate interest in advancing medical science. Imagine that you have been given a study protocol that has the approval of your institution and your peers. You anticipate that the results of the study may, in time, feed back in to help reduce the burden of disease. A large investment of time and effort has gone into the study. Some questions are asked, however, after a certain length of time, about the ethical probity of the study. A committee meets and recommends that the study continues. What would you do? Go against the authority of the protocol of the study and abandon it? Go against the findings of the committee, especially convened to address these issues, and abandon the study? Or consider that the balance of judgement means that the scientific merits of the study have been weighed against some ethical concerns, and have been vindicated, and continue? This is a close parallel to the situation of those involved in the Tuskegee study after the first recommendation that the study continue.

Tuskegee itself is an extremely complex and highly charged case, so I am using it more simply to think through abstractly about certain issues relating to moral authority in research ethics, and in what I say, I do not at all mean to distract from the wrongs suffered by those involved. Those conducting the study could perhaps have just abandoned it, or at least decided to quit and find another job. We cannot of course get back inside their minds. But it might be a mistake necessarily to think that, in such a situation, individuals are choosing between the moral and the non-moral, as is often portrayed. In sticking with the study protocol, they were sticking with one form of moral authority. Furthermore, a United States Public Health Service Commission looked into the study, found it acceptable and recommended that the study continue. A researcher then might well consider that it was wrong to discontinue. It begins to look less clear what one should do. If, with hindsight, we can say that the authority of this commission was inadequate or their judgement in this case should not have been followed, this may in part be because we were not there, did not directly experience the pull of this authority, and did not have to grapple with conscience. And we should consider carefully Milgram's findings of the commonly delayed reaction of conscience (Milgram 1974).

It is important too, to remember that the studies we are considering concern medical research, which aims to improve healthcare. Hence, in considering the ethics of continuing a study, it is not even a matter of weighing up morality versus "hard" science, the embodied concerns of humanity against the lure of brute fact. It is a matter of weighing human values on both sides of the equation.

Note well that the study took place over several decades, over which time there were some very significant developments: there were significant changes in attitude towards medical and research ethics, and also towards the rights and civil status of African Americans. During this time, significant developments in medical treatment occurred. During this time too, significant developments in the ethical regulation of research also occurred.

The long time-course of this study alone should give us pause for thought. Many of the studies involving genomics plan to retain data and samples for many years for possible future uses and reanalysis. It is humbling to wonder how studies embarked on today will look in fifty or so years' time, and to consider what contemporary currents of thought are affecting our views. Hopefully, the very fact that there are efforts to ensure the ethical conduct of research will mean that the passage of time judges us well. But, without forgetting the significant racial element in the conduct of the Tuskegee study, it should be recalled that at the time that the study commenced, it was common to conduct research without the standards of consent applicable now, and that a paternalistic attitude of doctors towards patients held sway.

Problems with Tuskagee arose not just from changing ethical attitudes and social mores, but with developments in science – the advent of antibiotics. This scientific development changed ethical stakes significantly. It is one thing to withhold information where there is no treatment; another where this means that effective treatment will not be sought. For any study enduring over such a long course of time, such significant developments are possible, even likely. There is indeed, a striking parallel with some of the major ethical problems facing genomics research: the issue of feedback of results. Many argue that it is fair enough not to report back results from genomics research of unknown significance or where there is no known or effective treatment. However, it is also pretty likely that over the course of time, treatments will develop and hence the question of feeding these back will arise. These points again will be discussed in greater detail in Chap. 8.

Meanwhile, a few conclusions may be drawn. We must guard always and eternally against moral complacency. Judgement of the actions of others must not preclude awareness that there, but for the grace of God and a hopefully vigilant moral awareness, go we. When we consider the ethical conduct of research, we must be mindful of the fact that we are making these judgements in a context where we are influenced by the authority of the institutional regulation of ethics, and that we make judgements in a specific geographical and historical context. This is not in itself to fall sway to a simplistic relativism; indeed, a recourse to relativism can lie easy on the conscience, since decisions can be justified with reference to local contexts, and inconsistencies compared to other times and places. It is the search for thoroughly grounded ethical responses that requires that we attempt to remain always conscious of the possibilities of our own local blindspots and prejudices.

7.4 Social Influences on Moral Action

There is a body of research in moral and social psychology which examines how and why we behave – and misbehave – morally. It is undoubtedly true that much more work needs to be done in this highly important area, but nonetheless we can still learn some useful lessons. Some research in psychology is snapped up by those with a penchant for "debunking" the aspirations of ethics – for example, interpreting research to show that we are not the agents of our own decisions, or that we are really at heart only interested in ourselves and our kin. But ethics is concerned not with the world as it is, but *as it might be*; but to get there, we need a full appreciation of how things stand and what obstacles might be in the way of any moral progress. We have noted the frequent disparity between stated aim and actual action; of the problems of weakness of will and self-deception. Often, human beings behave in ways that with hindsight, they regret. Understanding the sources of any poor behaviour is the first step to addressing it.

Here again, there is space only for a cursory overview of this field. We have already looked at spoilers of autonomous action and the influence of authority. What follows is a brief discussion of findings of particular salience to issues raised in this book. Those interested in a more thorough treatment of these issues will find a great deal of material: a good place to start is with Philip Zimbardo's book *The Lucifer Effect: How Good People Turn Evil* (2007), and its accompanying website http://www.lucifereffect.org. Do not be led to think from the title that this work has relevance only to occasions of great moral wrong doing. It also is greatly relevant to more ordinary acts of moral shoddiness and neglect.

7.4.1 Institutional Ethics: Administrative Evil

'The apocalyptic view of the world is one in which events do not recur' Ludwig Wittgenstein, *Culture and Value*, 1947 (1980).

Guy Adams and Danny Balfour's book, *Unmasking Administrative Evil* (2009) examines how moral wrongdoing can arise in the context of modern institutions. They discuss how a "technical rationality" used in the service of such institutions can mask unethical or even outright evil actions by undermining normative judgements and consigning ethical considerations into a box. They are concerned with how the creation of professional standards can lead to a lack of historical awareness and render engagement with ethical issues impotent. They argue that ethical standards can be applied in a formulaic way, leading to a procedure for deducing the "morally correct answer", and that ethical behaviour can then become little more than window dressing. They also argue that professionalism itself, by its conscious statements of norms and uniformity, can exacerbate tendencies to compliance in groups and organisations, a habit of compliance which can in turn help to foster a culture where "administrative evil" – decisions and actions with often unintended, but, nonetheless, disastrous effects for those affected – can thrive. Moreover, there is a dual intent in statements of professional ethics – not just to provide a set of formulae that dictate standards of conduct, but to foster public trust in the probity of the profession. This last intention can easily become at odds with the first as appearances and form gain ascendancy over substance. This was in fact precisely one of my worries about the inclusion of ELSI work within scientific projects.

They also warn that our contemporary ideology assumes autonomy and suggests that individuals will triumph over social forces, when there is a great deal of evidence to suggest that this is not always the case, as I have already argued. They point to the dangers of seeing evil events as one-off aberrations, since that precludes understanding the causes. Likewise, in genomics and medical research ethics, there is a great danger that the use of the highly extreme example of the holocaust may incline us to think that "we" could never do this (but then remember the whole point of the Milgram experiments). I would perhaps add to this the problem with using the language of evil: very often this leads to the dismissal of the relevance of such work on the grounds that the worse we could be dealing with would not be so bad as to class as "evil". But we need to look more inclusively at administrative wrong-doing.

Adams and Balfour point out that the abuses of Nazi Germany were only possible through the machinery of a vast and efficient bureaucracy; recall how the presence of professional ethical standards in Germany failed to prevent the horrors, many of which were meted out at the hands of the medical profession itself. They discuss how compliance with the demands of a large institution and its rules is easily ascertained, and discuss in detail a number of examples where systems led to disastrous failures with costs borne by individuals, and how the demands of instrumental and technical rationality can lead to individuals being viewed as expendable. These include the space shuttle disasters of Challenger and Columbia at the Marshall Space Flight Centre, NASA, the evacuation and relocation of all those of "Japanese ancestry" in the USA in the Second World War, and the tortures at Abu Ghraib of Iraqi prisoners by military personnel.

I first read Adams and Balfour's book at the suggestion of Philip Zimbardo, and long after I had myself grappled with the worries that ELSI work and ethical regulation in science and medicine might function simply as a box-ticking exercise. Their account of how professionalism and the institutional setting of standards can in fact act to mask underlying ethical problems fitted very exactly into my concerns. One of the major reasons why I am writing this book is to encourage deep, critical, and personal thought and action in relation to the whole conduct of scientific research; this is also a reason why I was at pains to stress that ethical concerns go far wider than simply compliance with ethics regulation or with appointing someone to do the ELSI work.

7.5 Hindrances to Ethical Action and Reasoning and What to Do About Them

7.5.1 Time

What has become known as the "Good Samaritan Sermon experiment" is often bandied about with a great deal of sidelong humour at the hapless subjects, which is unfortunate if it is allowed to act as a distraction to its lessons which could apply to us all: when we are in a hurry, we might not notice things around us which are of potential great moral relevance (Darley and Batson 1973). Other work too has provided an empirical demonstration of this truth: Time matters (Zimbardo 2008).

Briefly, in this experiment, theology students were asked to give a sermon, aptly on the topic of the Good Samaritan: the parable told by Jesus in answer to the question "who is my neighbour?" (The answer illustrated by the parable is: anyone who needs your help, even at some risk and cost to yourself, and even if they are from a social group with whom you are not on good terms.) The students had a short walk to where the audience waited for them: some were told that they had plenty of time, some that they were late and needed to hurry. On their way, a stooge sat slumped in an alley, as if in need of assistance. The findings of the experiment showed that those who thought they were in a hurry were far less likely to stop to see if the person in the alley needed any help, despite the fact that they perhaps should have been reminded of such a neighbourly duty by the content of the parable; the Samaritan himself stopped to help an injured man in a remote route populated by bandits.

This experiment is sometimes interpreted to show perhaps with a touch of schadenfreude, that studying theology does no particular moral good; more broadly and more fairly, that time pressures can make us neglect morality. The authors, however, draw a more nuanced conclusion, pointing out that those who hurried past were not simply neglecting a moral demand, but attending to another, since they had made an undertaking to others who were expecting them to fulfil this obligation at a particular time. Of special interest are reports of many of the subjects who said that it was only in retrospect that they recalled the slumped figure in the alley: the pressure of time may not, after all, have made them form a particular moral decision to ignore a fellow human in apparent need, but rather narrowed their attention so that they did not even consciously see him. (Recall our brief discussion of the importance of "moral vision" and attention in ethical reasoning in Sect. 4.3.)

This is of relevance to any moral decision making and action, especially where there are time pressures. We have looked at the frame of speed in genetics and genomics; there is common talk of an "urgent need" to attend to ethical issues. A major procedural issue in dealing with the ethical challenges in rapidly developing areas is indeed how much to rush to judgement and how much to think a bit more slowly. On a day to day level, meeting deadlines, dealing with heavy workloads, rushing to produce and to publish, can all divert our attention to these pressing goals and incline us to overlook things that merit our attention, even if those things are right under our very noses. Working conditions where there is constant time pressure and constant need to produce results may not produce the best ethical environment. Some of this pressure of time is hard to escape. It is perhaps ironic that some of it is however culturally and institutionally inflated by a number of factors, including drives to measure performance and to instil measures of accountability. Time pressures seem to have played a role amongst many influences that led to the erosion of moral sensibilities at Enron (McLean and Elkind 2004).

There is another connection with speed and morality: Milgram's finding that our moral conscience seems to suffer from jet lag. His subjects grappled with their consciences for some time before acting on them, if they indeed ever did. The "still, small voice" speaks, and we dither. This provides further evidential grounding for allowing the time to think things through ethically.

Concentrating on a goal ten minutes into the future can incline us to neglect something right under our noses. Concentrating on relatively short term goals can make us neglect longer term goals and values even if these are greater. In another famous illustration of a fundamental human failing, the Marshmallow experiment offered a group of children one marshmallow now; or two marshmallows (or jelly beans or Oreos) if they would wait a short time (Shoda et al. 1990). Years later, whether or not children waited for the greater reward proved a robust predictor of which children did well at numerous complex tasks involved in success in education and employment, as well as staying out of trouble with the law, as well as maintaining important personal relationships. The key seems to be the ability to delay gratification, to resist the delights of the present time for future goals. Whilst it is possible to look too much at the future at the expense of "time off" and the simple enjoyments necessary for a rounded life, a time perspective that focuses too much on the present may make us seek immediate rewards at the expense of more enduring things.

However, it must not be thought simplistically that behaving ethically consists in looking longer term to the future. Looking to the future can make us neglect the ethical demands of the present. A stunning example of this is Albert Speer's theory of "ruin value" in architecture: buildings he designed for Hitler were purposely intended to look good as ruins so that thousands of years hence, the Third Reich would be judged accordingly as a great empire; yet he neglected the moral atrocities happening in the present (Speer 1970). Flexibility in orientation and perspective – judging relative importance of potentially competing ethical demands – is perhaps of more importance than having one particular time perspective.

What is the relevance of this to the conduct of scientific research? One big problem with the practice of science is not only the time pressure of speed, but the urgency of this – what hangs on this. Short term funding of research projects, and short term employment contracts not only channel our attention to particular periods of time, their power to do this is immense because getting the next grant, and getting the next employment contract, are the *sine qua non* of the whole enterprise. It is widely accepted that the short term nature of the election cycle has the drawback that politicians focus on short term goals that they can accomplish preferably with visible results within their period in office. This general problem applies too to the system of research funding. This is not to say it is insuperable; many who work within research are motivated by very long term goals and norms, including concern for the future. Indeed, contributions to the future advance of science, a wish to leave a lasting legacy in one's work and to serve the common good are also major human motivators.

Time, Ethics, and Reflection

The moral system of utilitarianism aims to produce the best overall consequences for everyone. One criticism of the theory is that this would mean that on every occasion, an agent must weigh up a vast amount of relevant information and calculate the most likely outcome of all the possible options for action. This would involve lightning fast calculations. Many have suggested a solution involving a two-tier system of ethical judgements. This involves acting on individual occasion when fast judgement is called for on the basis of rules of thumb drawn up to make rough and ready decisions, and then, on reflection at leisure, considering and perhaps revising the details of these rules of thumb on the basis of experience and consideration.

A two-tier system of reflection can likewise help to address the problem of decisions that need to be made under some time pressure, even if one is not a utilitarian. In the context of research ethics, this would require that active consideration is given to allowing time for reflection, and, vitally, feeding back into this system of reflection any lessons learned from the application of rules of ethics to particular cases. A crucial question that arises is the relation between the two levels, and who gets to input into the higher level. A related

(continued)

and crucial question is collecting the evidence needed to feed into the higher level. And a further question concerns what to do about realisations that things are not ideal and have not been as they should be (Hare 1981; Crisp 1997).

7.5.2 *Dehumanisation*

There is a plethora of work that indicates that, under certain conditions, moral restraints are liable to break down; any conditions that incline towards dehumanisation are likely to encourage maltreatment. Conditions of dehumanisation are any that make it less possible to view people as individuals, with identities, values, and places in a social world where others also value them.

What makes for dehumanisation? Experimental work has demonstrated effects of dehumanisation and its opposite under a number of conditions (Haslam et al. 2005). Anonymised subjects without names, or with numbers in place of names; those not visible, or with masks or otherwise with their face shielded; calling individuals or groups by derogatory names (such as often happens in warfare, in abusive terms for the enemy, or in "euphemisms" which hide the gruesome details of actions, such as referring to the dead and injured as "collateral damage"). Simply telling people that others are socially inferior can have startling effects; in the well-known "blue eyes brown eyes" experiments, school children who were simply told that those with a certain eye colour were inferior quickly started treating them with disdain (Peters 1971); in the Stanford Prison Experiment, simply assigning healthy students at random to the role of prisoner or of guard unleashed abusive behaviour on the part of most of the "guards" with startling rapidity (Zimbardo 2007).

What particular relevance do such findings have to genomics research? There is a possible tension in that the requirements of confidentiality require that data is anonymised. Now, anonymisation is not in itself dehumanisation. It is possible to treat people with great humanity and sensitivity even without knowing their names. But could there possibly be any ways in which the routine anonymisation of research data reproduce the ethical problems of dehumanisation that anonymisation of subjects often creates? This is an interesting and difficult question, one that relates to questions of the distance or proximity between researcher and subject that are discussed in more detail when we examine researcher responsibilities in Chap. 9. One response to this might be to create opportunities for reminders of the humanity of research subjects, for example through representation on governing bodies or ethics committees, or for communication with subjects, for example through newsletters about research findings. Many research projects do precisely this. It would be interesting to bear in mind the question of whether ethical decisions affecting research participants are affected by their relative invisibility.

7.5.3 Social Conformity

The hubris of humans was shaken on the discovery that we share so much of our genome with simple animals like worms, that we have no more genes than some very primitive organisms. But it would be better ethically if we paid more attention to the worrying ways in which we are so much like sheep.

A great deal of well corroborated research in social psychology has shown the power of social conformity to sway even the judgements of our own eyes. The famous Asch experiments (variations of which have been repeated many times) showed that faced with social pressure, subjects could easily be induced to say that a line was longer than another line, even when the opposite was true (Asch 1951, 1955). One might speculate that in cases of finer or complex judgement which ethics almost always involves, such social pressure to conform might be stronger still. It would perhaps have been a factor in the decision of those involved in Tuskegee to continue the research – everyone else around them doing it was continuing, weren't they?

The Milgram experiments included variations that showed that where the "teacher" had a role model of someone else who refused to continue to shock, far more also refused to continue. This in itself may be an example of social pressure. Or it could be explained in terms of legitimating dissent.

A lesson to be learned would be to take steps to ensure a variety of viewpoints. As noted above, the very fact of institutional and international collaboration in research and the fact that those of different disciplinary backgrounds are coming together in research should make this child's play, so long as institutional space is given to it.

How might this be achieved in practice? Other work in social psychology shows the importance of encouraging self-awareness in moral reflection. The simple placing of mirrors in phone booths acts to discourage vandalism: the sight of catching oneself in the act can be enough to bring you up short (Zimbardo 2007). As seen in our earlier discussion of Nathan and David, a third party can play this role. The necessity of explaining oneself to someone else can provide such an opportunity. The inclusion of dialogue with members of the public or representatives of research subject or patient support groups can likewise assist with moral self-reflection. Zimbardo discusses the benefits of belonging to more than one social group, as chances then arise for the differing values of the different groups to challenge and stretch each other (2007). The value of activity outside of one's research work, and the possibility of belonging to more than one professional group, can be of ethical value. The possibility of dialogue and dissent must be real, however. William James in his book *The Will to Believe* discusses how a belief may be a "dead" or a "live" option (James 1897). A belief is only a live option if it is one you might actually decide to take on. For most people, the belief that the world is flat is a not a live option. Some moral beliefs may not be ones we would take seriously. Perhaps a more nuanced category could be drawn, whereupon some beliefs are less likely to be taken seriously. We need to ensure that we are not writing off possibly valuable viewpoints in their entirety. Allowing an arena for social disagreement, active debate and the consideration of even unpopular viewpoints needs to be more than a token nod at "doing ethics".

7.5.4 Moral Disengagement

There have been many studies on how a person can switch from acts of cruelty to acts of compassion, and how people can act seemingly completely counter to their professed moral views (Bandura et al. 1996; Zimbardo 2007). Some studies have concerned themselves with morally vivid cases, such as the behaviour of those involved in state executions which may seem a far cry to the issues involved in genomics research, some with more mundane issues. Nonetheless, the ease with which actions can be pigeonholed off into realms which the relevant moral codes seemingly cannot touch should be noted carefully: yet again, this is empirical confirmation of phenomena akin to self-deception which has long troubled philosophers, moralists and theologians.

7.5.5 The Diffusion of Responsibility

We have already considered in some detail how notions that someone else is responsible are a major mechanism for moral disengagement, in the examination of the Milgram experiments. People can take on many different social roles, and fulfilling the requirements of these can be used as a justification for behaviour: "I was only doing my job" (Arendt 1994). Fragmentation of social roles can potentially mean that overall responsibility for outcomes is hidden. Passing the buck is also an ever lurking trap: the temptation is often to ask "who is responsible for this" and finding that it is not oneself, to assume that therefore, *it is someone else* (Darley and Latane 1968; Moriarty 1975).

Sometimes yes, but sometimes no. In his analysis of situations in which ordinary people are led to perform unethical actions, Zimbardo argues that it is necessary to look at more than just the individuals involved (2007). We need also to look at the social situation; and we also need to look at systemic factors operating. We are often inclined to make what psychologists call the "Fundamental Attribution Error", and conclude that features of a situation can be ascribed to characteristics of the individual agents involved. But often, better or additional explanations come from considering the social settings and wider institutional and systemic factors. We will consider the vital issue of responsibility at greater length in the following chapter.

7.5.6 *Whistleblowers*

Problems such as those of social conformity, the phenomenon of moral disengagement, and the diffusion of responsibility, all underline the importance of whistleblowers and the need for their protection. This needs to be institutionally recognised. Even where steps have been taken to protect whistleblowers legally, it is still being found that in some cases whistleblowers may have problems such as difficulty in finding future employment. For examples, there have been allegations that whistleblowers within the NHS have suffered considerably for their actions (Cohen 2011). This underlines the potential gulf between the machinery of an ethical institution and its actual accomplishment.

Philip Zimbardo's Ten Step Programme to Resist Unwanted Influence
This suggested list for the toolkit against poor decisions and unwarranted social influence has been developed by Zimbardo based upon extensive work of his own and others in social psychology.

ONE: **"I made a mistake!"**

TWO: **"I am mindful"**

THREE: **"I am responsible"**

FOUR: **"I am ME, the best I can be"**

FIVE: **"I respect just authority but rebel against unjust authority"**

SIX: **"I want group acceptance, but value my independence"**

SEVEN: **"I will be more frame-vigilant"**

EIGHT: **"I will balance my time-perspective"**

NINE: **"I will not sacrifice personal or civic freedoms for the illusion of security"**

TEN: **"I can oppose unjust systems"**

Summarised from Philip Zimbardo, *The Lucifer Effect*, Chap. 16, "Resisting situational influence and celebrating heroism" (2007).

7.6 Conclusions

Applied ethics broadly and research ethics which is our particular subject, have in general paid attention to the issue of what are the best ethical decisions, and much less to the issue of how these decisions are made in practice, and to the question of moral motivation. In considering ethical issues, we need to look at the social, economic and institutional context of science, examining both its internal operations and its relation to the wider society. We can also learn a lot by considering findings about individual and social psychology. It is especially important to consider lines of authority and the influences of power upon both our actions and our beliefs. Particular problems may arise when there is institutional hierarchy and pressures of time, money, and performance. Much research has shown how these are affected, often to deleterious effects, by pressures such as social conformity, and has demonstrated the ease with which people may disengage morally and may dehumanise each other. But conversely, this also shows us the traps to watch out for and suggests ways to avoid such failings. Putting in place a code of ethics is never going to be in itself adequate. Attention to the social and psychological reality of belief and action is of overwhelming importance in the goal of the ethical regulation and conduct of research.

7.7 Study Questions

1. A standard way of dehumanising people we are dealing with is to fail to use their names, for example to give them numbers instead. There are good reasons in many contexts to keep a person's identity private, and this may be achieved by using numbers only and not names, as when anonymisation takes place with research data. Are there any dangers that such processes of anonymisation may lead to the failure to recognise the humanity of those concerned? What parallels or disanalogies can you see in these situations? What practical steps can research projects take to avoid any possibility of such dehumanisation?
2. Consider the institution within which you are currently working or studying (or a past example if you are not currently a member of any formal organisation). If you were concerned about the actions or attitudes of someone above you in the institutional hierarchy, what would you do? Do you know what channels you might use to express your concerns? What costs might there be to you of doing so? How would you weigh up whether the possible gains from speaking up might outweigh the costs to yourself?
3. There has rightly been concern that commercial organisations may have vested financial interests that may bias their judgement in various contexts. What sorts of interests, financial or otherwise, might affect the judgement of those working in scientific research? What steps could be taken to mitigate these factors?

7.8 Coming Up Next

We have looked in some detail at individual autonomy, at problems with this concept, especially within the context of genomics. A very closely aligned concept is that of respect for persons. This is often used in a looser and broader way; it is also an important concept used to counter ways in which individuals or groups are dehumanised by others. As we have seen, autonomy can be understood in many different ways although is widely valued, but also recognised on many occasions to be in tension with community values which may override considerations of individual autonomy. Respect for persons is in many ways then a more powerfully persuasive term, as it is harder to try to argue that it should be overturned in favour of other values. The next chapter delves into the notion of respect for persons, how it is understood, what role it may have in ethics in general and in genomics research in particular. The chapter also examines how this concept is applied to a particular debate in research ethics, and looks in detail at how conceptual confusion and weaknesses in chains of argument can lead to unwarranted conclusions.

References

Adams G, Balfour D (2009) Unmasking administrative evil, 3rd edn. Sage, London
Angell M (2005) The truth about drug companies: how they deceive us and what to do about it. Random House, New York
Arendt H (1994) Eichmann in Jerusalem: a report on the banality of evil. Penguin, London
Asch SE (1951) Studies of independence and conformity: a minority of one against a unanimous majority. Psychol Monogr 70:416
Asch SE (1955) Opinions and social pressure. Scientific American November 1955
Bandura A, Barbaranelli C, Capara GV, Pastorelli C (1996) Mechanisms of moral disengagement in the exercise of moral agency. J Pers Soc Psychol 71(2):364–374
Centre for Genetics and Society. http://www.geneticsandsociety.org/
Centre for Disease Control U.S. Public Health Service Syphilis Study at Tuskegee. http://www.cdc.gov/tuskegee/timeline.htm. Accessed July 2011
Cohen N (2011) Let the law save whistleblowers, not silence them. The Observer, July 10 2011
CONSORT Consolidated standards of reporting of trials. http://www.consort-statement.org/
Crisp R (1997) Mill on utilitarianism. Routledge philosophy guidebooks. Routledge, London
Curran WJ (1973) The Tuskegee syphilis study. New Engl J Med 289(14):730–731. doi: doi:10.1056/NEJM197310042891406
Darley JM, Batson CD (1973) "From Jerusalem to Jericho": a study of situational and dispositional variables in helping behaviour. J Pers Soc Psychol 27(1):100–108
Darley JM, Latane B (1968) Bystander intervention in emergencies: diffusion of responsibility. J Pers Soc Psychol 8(4):377–383
Davies K (2001) Sequence: inside the race for the human genome. Phoenix, London
ESRC Centre for Economic and Social Aspects of Genomics. http://www.genomicsnetwork.ac.uk/cesagen/. Accessed July 2011
Foundation for genomics and population health. http://www.phgfoundation.org/
Gene Watch UK. http://www.genewatch.org/
Genetic Alliance. http://www.geneticalliance.org/
Hare RM (1981) Moral thinking: its levels, method and point. Oxford University Press, Oxford
Haslam N, Bain P, Douge L, Lee M, Bastian B (2005) More human than you: attributing humanness to self and others. J Pers Soc Psychol 89(6):937–950
House of Lords Select Committee on Science and Technology. http://www.parliament.uk/hlscience. Accessed July 2011
Human Fertilisation and Embryology Authority.
Human Genetics Commission. http://www.hgc.gov.uk/
James W (1897) The will to believe: and other essays in popular psychology. Longmans, Green, New York
Jones JH (1981) Bad blood: the Tuskegee syphilis experiment. Free Press, New York
Mayes T (2002) Restraint or revelation: free speech and privacy in a confessional age. Spiked, London

McLean B, Elkind P (2004) The smartest guys in the room: the amazing rise and scandalous fall of Enron. Penguin, London
Medawar C (1992) Power and dependence: social audit on the safety of medicines. Social Audit Ltd., London
Milgram S (1974) Obedience to authority. Harper Collins, New York
Moriarty T (1975) Crime, commitment, and the responsive bystander: two field experiments. J Pers Soc Psychol 31(2):370–376
Nuffield Council on Bioethics. http://www.nuffieldbioethics.org/. Accessed July 2011
Peters W (1971) A class divided. Yale University Press, New Haven, CT
Resnik DB (1998) The ethics of science: an introduction. Routledge, London
Resnik DB (2007) The price of truth: how money affects the norms of science. Oxford University Press, Oxford
Shoda Y, Mischel W, Peake PK (1990) Predictive adolescent cognitive and self-regulatory competencies from preschool delay of gratification. Dev Psychol 26:978–986
Speer A (1970) Inside the Third Reich. Weidenfeld and Nicolson, London
Unmasking adminsitrative evil. http://unmaskingadministrativeevil.wordpress.com/. Accessed July 2011
Wittgenstein L (1980) Culture and value (trans: Winch P). Basil Blackwell, Oxford
Zimbardo P (2007) The Lucifer effect: how good people turn evil. Rider, London
Zimbardo P (2008) The time paradox: a new psychology of time. Rider, London

Chapter 8
Respect for Persons in Research and in Genetics

Respect for persons affirms the equal value, dignity and moral rights of each individual. Each individual is entitled to lead a life in which genetic characteristics will not be the basis of unjust discrimination or unfair and inhuman treatment.
Inside Information: Balancing Interests in the Use of Personal Genetic Data
Human Genetics Commission (2002)

Abstract This chapter examines a central concept in ethics, that of respect for persons. This notion is widely used but understood in loose, varying, and sometimes inconsistent ways. Common ways of understanding the notion of respect for persons are explored and philosophical discussion of respect for persons is introduced, including the account given by Immanuel Kant. The scope of any imperative to show respect for persons is examined, together with its relation to personal information. The debate about the feedback of findings in genomics research is used as an illustration of how the notion of respect for persons has been used to produce practical ethical conclusions. However, a close examination of the literature shows how varying notions of respect are elided and how unwarranted conclusions may be drawn. Conclusions drawn include the need for careful elaboration of the notion of respect for persons, and the need to examine the evidential and argumentative support for a position.

8.1 Study Question

Study questions normally go at the end of chapters. For this chapter, for reasons which will soon be clear, here is one to start you off.

Spend a few minutes jotting down what the term "respect for persons" means to you.

If you are part of a class or know others who are reading this book, compare your answer with theirs.

P. Boddington, *Ethical Challenges in Genomics Research*,
DOI 10.1007/978-3-642-23699-0_8,

8.2 Introduction

This chapter looks at how the notion of respect for persons has been used and misused in some ethical discussions in genomics. The idea of respect for persons has come to inhabit an important role in moral and political philosophy, and in more everyday thinking about ethics (Dillon 2010). The imperative to demonstrate respect for those taking part in research is a central concern of research ethics; the notion of respect more generally is a vital plank of our ethical world. However, there are many different accounts of just what it is to exhibit respect for someone, and many different avenues of disagreement about the term. There are accounts of respect for persons that have received extensive philosophical elaboration, and there are intuitively plausible, everyday uses of the term. There is hence ample scope for confusion and for masked disagreement.

For example: respect may refer to an evaluation of someone's capacities in a certain arena, as one may respect the playing of a great pianist; it may refer to the recognition that a person or an object or force may represent a potential obstacle that needs to be taken into account, as for instance one may respect the power of the sea; it may refer to the institutional trappings of respect, as when one shows respect for the monarchy or the governing body of a university by conforming to its edicts and protocols; or it may refer to a core notion of respect that is said to be owed to all human beings, or all persons. This latter notion of respect is often claimed to hold regardless of rank or individual variations in talents or capacities, but to derive from certain core capacities or properties belonging to all who merit such moral respect.

It is this latter sense that centrally concerns us here. But even in this sense, there is room for considerable disagreement. Difficulties in argumentation involving respect for persons are legion partly because the basic notion of respect is so ubiquitous in ethics, yet plays such a number of diverse roles that these are easily conflated and provide countless opportunities for confusion and vagueness. This is confounded by the understandable lure of the notion of respect, and the difficulty of naysaying it. For who is against respect? But we can think we have agreement, and may be put off questioning in case it looks like we are "disrespectful", but this can mask and even perpetuate serious disagreement.

This chapter will also illustrate certain key methodological issues in ethical argument. These include the importance of examining the history and analysis of a concept, the importance of clarity about detail, and the importance of tracing back how ideas are shared and cited through a continuing debate. Additionally, we look briefly at how problems in ethical arguments can be created through failing to pay sufficient attention to the details of the relevant science and scientific practice. The example of debates about feedback of research findings to research subjects is chosen as a vehicle not just for its inherent interest but also because it shows good illustration of the perils of the use and misuse of the notion of respect for persons.

8.3 Varieties of Respect: Divergent Understandings of the Concept of Respect for Persons

The philosophical literature encompasses a range of different accounts of what is meant by "respect for persons" (Downie and Telfer 1969). Perhaps as importantly for the readers of this book there are substantially different notions of what "respect for persons" involves which are held more generally. In 2010, I canvassed views about what this concept meant from a group of about 40 people attending an ethics workshop at a conference on data sharing in genomics (held at St Hugh's College, Oxford, September 2010). I am grateful to those involved who contributed a wide-ranging and often detailed set of responses to the question "what does respect for persons involve?" Responses varied widely.

Some responses were abstract expressions of respect; for example, referring to notions of "fundamental humanity", "autonomy", "inviolableness" and "privacy", or "dignity", to treating people "as human beings not as objects" and "as independent human beings".

Some responses developed a negative expression of respect for persons as "noninterference". Some referred to freedom of action, some to freedom of decision or choice, and some to freedom of value and opinion. Some alluded to an underlying reason: ("assuming they are best suited to determine their own ends"). Some restricted such freedoms "within agreed-upon limits", in contrast to many others which implied that people should be allowed "to do whatever they please".

Some responses gave a stronger, positive conception of respect for persons as enabling growth and action. Some of these responses referred to the need for adequate information to make decisions (indeed decision making without the necessary information may seem to be a rather empty freedom). But some responses referred to far more substantial notions: "allow personal flourishing" and "a desire (and duty) to assist them in developing themselves and minimising vulnerability". These accounts impose quite weighty obligations on others, and also rely implicitly on particular accounts of what is required for human flourishing and well-being, where there is great scope for disagreement.

Some responses did indeed explicitly make substantive ethical claims as they incorporated rights, equality, and other basic moral values into the notion of respect for persons. But these notions are not at all equivalent to each other; indeed, there are very different ways of understanding the notion of equality, which could refer to, for example, equality of opportunity or equality of distribution of social and economic goods and services.

Rather different substantive issues are raised by responses concerned with the consequences of our actions, sometimes specifically referring to the avoidance of harm. This is interesting particularly as the notion of respect for persons is often drawn upon as a device to rein in the excesses of consequentialism, given the well-discussed problems of consequentialism in failing to recognise the separation of persons (see Sect. 6.2.5).

Indeed, several responses referred to some notion of the separation of persons in their explication of respect for persons, for example, "give value to their way of thinking and not impose your own views above theirs" and "help to safeguard space for everybody to pursue their aims".

A related group of responses expressed a further and equally important view, that respect for persons implies the absence of exploitation and that no person is to be used for another person's purpose. These included "not using them", "do not exploit someone for my own means", and "be seen as ends, not as means". This is of course a response explicitly opposed to a strict consequentialism which is concerned exclusively with the ends of our actions.

Several responses thus mentioned transparency: "being transparent in our dealings with people". Other responses called upon notions of reciprocity: "treat others like we would like to be treated", "do unto others what you would like done to you", and "compromise".

Interestingly, the quality of relationships between people, of which reciprocity is a part, was also strongly emphasised in many responses which cited understanding and communication between people as a component of respect for persons. Responses included: "empathy", "listening", "a willingness to understand what he/she values", "first we need to know the person", "communication", "consultations when in doubt", "being sensitive to the concerns of others", and "try to understand *why* they are behaving in a particular manner". It is also interesting that this last response added: "but not necessarily accept their behaviour or not trying to stop what they are doing". Dangers still exist, however, which is exemplified perhaps in this response: "Interacting with them solely for the purpose of bring(ing) them to the table in matters that affect them. What does this mean. One-time consent? Tokenism?"

Several responses discussed the relationship of the individual to the wider community. Respect for persons is generally thought of as an individual right, but one respondent wrote: "manifested in both reactions to individuals and groups. Could be a collective right (minority group)". And another saw tensions between group and individual respect: "With whom to connect – representative sample – who has a right be to spokesperson?" A substantive response linked respect for individuals to "Granting them the privilege of contributing to the common social good".

8.4 Background: The Debate About Feedback of Results

There has been, especially in recent years, a lively debate about whether, when and how, results of research studies should be returned to participants. There is now broad consensus that general findings of research should be available to the public and in particular to those who took part in the research, although making such access easy is still far from universal practice (see Chap. 10 for further discussion of the value of open access to scientific knowledge). The area of contention concerns the return of individual-level findings. These include incidental findings – those extraneous to the subject of research – as well as those under direct investigation.

Different areas of research pose different challenges. For example, using imaging techniques it is sometimes possible for researchers to discover unexpected findings of great significance to an individual's health, such as a tumour. In such cases, most pressingly where effective treatment is available, there is common agreement that feedback should be given. However, in the case of much genomics research, the issues are complex and less straightforward. In some cases of research, a practitioner is there in the room with a subject on the discovery of an unexpected anomaly on an image; in much genomics research, there are various highly relevant differences. Data may be analysed at considerable removal of time and distance from the participants, in an anonymised and aggregated form, and its significance may be more difficult to ascertain. Hence, there is a more complex debate about whether or not certain results should be fed back to individuals.

An examination of the contemporary debate demonstrates that one key concept used, especially by those who advocate the return of results, is that of respect for persons, which is argued by many to mandate the return of results. We shall examine carefully whether or not this claim is convincing. We will see how "respect for persons" is often used vaguely and without sufficient anchor for the claims that rest upon it.

Immanuel Kant and Respect for Persons

There are many different philosophical accounts of the notion of respect for persons but the granddaddy of them all is the account given by Immanuel Kant in his *Groundwork for the Metaphysics of Morals* (1948). Virtually all subsequent accounts build upon Kant's account in some way. His account is of a concept owed to all of humanity, insofar as they are rational agents. (It should be noted that this notion of rationality is simply a general capacity for reason, and does not at all imply, for Kant, that those with higher capacities to reason merit any greater respect.) A rational being is one who can recognise the ground of his or her actions, and act in accordance with certain laws or "maxims" of action.

For Kant, respect for persons is an expression of the dignity or ultimate worth of human beings. This is contrasted with conditional worth, which a person might have, for instance insofar as they possessed a useful skill, or had a certain rank in society. Respect for persons, for their dignity, and their ultimate moral worth and value is in contrast unconditional. It is the universal and unconditional nature of respect for persons that makes it such a powerful expression of the value of individuals.

Kant produced various formulae by which he expressed the universal law of morality. We have already met the second formulation:

> *Act in such a way that you treat humanity, whether in your own person or the person of any other, never simply as a means but always at the same time as an end*

(continued)

This simple formula is at once intuitively plausible yet deeply mysterious. It is vital to note that Kant ruled out treating another person as *simply* a means. It is a part and parcel of everyday life that we treat people as means; to pass the salt, to drive us to town, to answer a simply query, to milk the cows, and to deliver the milk. What we should be doing *at the same time* is to recognise the inherent worth of each person as a member of humanity, who has dignity and not just price – something that can be traded off for another thing of equivalent value. The elusive part of this attractive idea is just how exactly we treat someone as an "end in themselves".

A weighty question then is what exactly we have to do to demonstrate such respect for the humanity of another. This is an area of great dispute that has significant implications for what moral conclusions one might draw. Respect for persons may be seen as a negative duty, a constraint on action. We must not, for instance, lie, cheat, steal, or kill. For others, respect for persons implies positive duties. There are various putative candidates for such duties. These include: to protect others from erroneous ideas that might lead to their harm, to try to see the world from their point of view, and to help them to achieve their (morally acceptable) desires.

A crucial distinction is between views which hold that there is some objective grounding to how we must act in treating other persons with respect and views which tend to a subjective approach, that treating someone with respect is centred on how they would like to be treated.

8.5 The Use and Misuse of the Notion of "Respect for Persons" in a Current Debate in Genomics Research Ethics

This section traces back the use of the notion of respect for persons, and cognate terms, through the current debate about return of research results especially in genomics research. The aims of this section are to demonstrate how the notion of respect for persons may be used in various ways; how related terms such as "respect" "respect for persons" and "respect for participants" might easily be elided in surveying different papers on a particular debate; and how the concept of respect for persons may be used as if it grounds particular conclusions, when those conclusions may stand in need of much more rigorous justification. Conclusions may be drawn too swiftly both in the papers and by readers. I would like to add that this analysis aims not to snipe at any of the particular authors used here in illustration: there are many examples I could have chosen. (It is also very likely the case that in many instances where ethical positions are put forward with inadequate grounding, or concepts used with insufficient elucidation, the authors could have supplied this grounding but were prevented from doing so by journal word limits.)

Although many express scepticism on various grounds, it is fair to say that there is a growing body of opinion that research results should be fed back to subjects. Much of this literature refers back to certain key sources and hence the use of notions of respect, and their alleged support for the return of research results to individuals, can be readily traced back through the literature. One of the first sources cited in this recent debate is Partridge and Winer (2002). The extract given below makes oblique reference to some classic notion of respect for persons, drawing substantive conclusions from it, without clearly referring to the range of views that exist concerning what we exactly owe to those who merit such respect.

"In general, the aim of clinical research should be to treat each individual participant with the utmost respect, as an end in himself or herself, and as a partner in research. In support of this approach, results should be shared, as doing so could be considered the correct course of action when working toward a common goal with a partner".

This is one beginning of a thread of ideas in the literature on feedback of genomics research results, unfortunately a beginning premised on a chimera of poorly articulated ethical ideas. The wording and the context strongly suggest that there is some allusion here to a Kantian notion of respect for persons. But what it is to treat a person with the "*utmost* respect" is not explained or referenced; this might be a way of expressing the notion that respect for persons is unconditional and equal for all persons. There is a danger however that "utmost" respect may be taken to imply that every possible step needs to be taken to mark respect. Partridge refers to the notion of an "end in himself or herself"; however, as we have seen, for some, respect for persons involves simply constraints on action, while for others, positive duties of various disputed sorts.

The step from treating someone as an end in him or herself to treating them as a partner in research is not adequately explained. Partnerships are many and various, and, for many good reasons, not all are equal (and lack of equality may be perfectly compatible with respect – or with a different notion of equality). Whilst no one would deny that participants in research should be treated with respect, the notion of treating them with "utmost" respect is not one that has any clear precedence in any identified philosophical literature and, moreover, one that potentially opens the door to whatever escalating demands that may be made on behalf of participants.

Equality

The idea of treating someone with respect is often closely aligned to the idea that people should be treated with equality. Any evidence of unequal treatment is thus sometimes taken as evidence of lack of respect. But what is equality? Does all inequality demonstrate lack of respect? The answer is that there are different notions of what constitutes equality, differences in treatment may be compatible with equality, and some sorts of inequality may be quite compatible with respect for persons.

If I wish to treat my two children equally, I could divide a loaf of bread into two equal halves and give them half each; or I could divide it unequally and

(continued)

give the larger portion to my 6-ft 2-in.-tall son and the smaller portion to my 5-ft-tall much younger daughter who has significantly lower calorie requirements, hence giving them each as much as they need; or I could say to both, "there's a loaf in the bread bin, you can both help yourselves"; or, suppose one child is gluten intolerant, I could give them equal amounts of food, but of different types – one gets rice cakes and is explicitly denied the bread. If it is beer, I could say I treat them equally yet treat them differently, in allowing my adult son to have beer now, but not my daughter, because once she is 18 she can have some too.

Not only are all of these compatible with different notions of equality, they are all compatible with some substantial notion of treating them with respect as well.

Equality of outcome gives them both half a loaf
Equality of desert gives them their individual food needs (and there is a variation which premises this on merit rather than on need)
Equality of opportunity allows them equal chance to get the bread
Treating different people differently because of morally relevant differences is not in itself unequal treatment – the access of both to beer is decided under the same, equal, and fairly applied rules

Another frequently cited source for the idea that return of results is required by an ethical notion of respect for persons is Fernandez et al. (2003). They claim:

"Respect for persons requires that choices made by individuals who are capable of making decisions for themselves be accorded high regard, and this is embodied in the concept of free and informed consent".

This definition equates respect for persons with free choices. In Kantian notions of respect for persons, it is only choices that persons make that are true to our natures as rational agents, and only choices that could be willed to become universal laws of nature, that need be respected. The formulation given by Fernandez et al. is more akin to an economist's model of preference satisfaction, and fits with a much broader version of respect for persons as autonomous choosers. (See the discussion of autonomy in Chap. 10.)

Moreover, doubtless aware of the problems that are quickly encountered in allowing individuals free choices, these authors only say that these choices given "high regard", which is yet a long way off the conclusion that they lead to any requirement of return of results. There is of course an inherent vagueness in claiming that any such choices should be "accorded high regard" as it is quite possible to give "high regard" to choices which are then overridden; choices of one person may, and often do, clash with the choices of others.

They argue that: "The act of offering a summary of the research results avoids treating persons solely as a means to an end". This introduces another formulation of the notion of respect for persons, but implicitly trades upon a glaring fallacy which has led to frequent misapplications of Kant. The Kantian notion that we should not treat people solely as a means sometimes leads to the thought that to avoid disrespect what we need to do is to *add* something to how we treat people. Even on this approach, there are obviously various ways of doing this. Offering summary research results would be a means of avoiding treating persons solely as means to an end, but is far from the only way of doing this. Simply saying "please" and "thank you" and explaining what you want, or offering a financial reward, for example, are also ways of avoiding treating persons solely as a means.

Much more importantly, this was not at all what Kant had in mind. As we have seen, no formulation of respect for persons rules out treating people as means – or else everyday human interactions would break down – but merely rules out treating people *solely* as means to an end. The wording has confused many into thinking that we need to do something extra to avoid disrespect. However, it is the nature of our treatment of others, not something sprinkled on top, which constitutes respectful treatment. If I am treating you with disrespect, I do not make it any better by giving you a free gift or saying "thanks". (I certainly cannot make it respectful by simply claiming that I have some respectful "attitude": a common abuse of notions of respect is to be heard in when someone says "no offence, but . . ." The "no offence" operator is *in itself* impotent to create respect where there is none.)

Treating someone as a means to an end would involve such things as manipulating them, failing to explain what was really going on, riding roughshod over their basic needs, and so on: Fernandez is on stronger grounds to refer to the notion of informed consent as a device for ruling out such moral skulduggery, and there is arguably then no necessity to offer anything additional to this to avoid treating people merely as a means. If the researcher has explained fully what the research involves, has obtained freely given consent, is being open and honest, and is doing research with ends that those taking part can endorse, there is so far no reason to think it not entirely compatible with respect for persons. If someone freely agrees to help you with your legitimate ends, you are not treating them solely as a means: you have respected their humanity, their rationality, by explaining what is going on and letting them freely choose.

Shalowitz and Miller's paper on disclosing individual results of clinical research is another oft-cited source for the notion that respect for persons grounds ethical obligations to return results (2005).

"Respect for persons is a basic ethical principle that gives rise to obligations regarding how competent adults should be treated. As a minimum, this principle of respect for persons prohibits treating persons as mere means to ends. However, additional specific obligations of respect derive from interpersonal or institutional relationships between persons."

"In biomedical research, respect for persons includes investigators' obligations not to coerce or deceive research participants and to obtain informed consent. Respect for persons, including respect for participants' self-determination and a recognition of their integral role in research, underlies investigators' responsibilities to make aggregate results available to participants."

This account takes off from a Kantian notion of respect for persons and then adds to this the idea that there are additional obligations of respect deriving from specific concrete relationships between individuals. This is surely the case. However, this adds not only a substantially different notion of respect, but additionally a realm of uncertainty as these will for sure be disputed and dependent upon particular arrangements. Even a slightly more detailed reference to the range of diverse accounts of respect for persons, as well as the different dimensions of respect and their concrete application, could have made this clearer. Without such clarity, Shalowitz and Miller step very swiftly straight from obligations not to coerce, deceive, or conduct studies without the knowledge and understanding of participants – which can all be seen as negative duties not to go against what a participant might want, and which all might be derived without problem from a Kantian notion of the respect owed to all of humanity – to positive duties to provide participants with aggregate results.

Negative and Positive Duties

A simple but important distinction.

Negative duties are duties to refrain from particular actions: such as wrongful killing, stealing, and lying.

Positive duties are duties to perform certain actions, for example, duties to render aid those in need.

Negative duties are often held to be binding on us at all times. It is easy to see that performing positive duties takes time and resources. Thus, there may be some scope for determining when and how exactly we fulfil these.

Negative duties not to treat a person with disrespect are generally held to be much stronger than positive duties towards them. Whether or not providing aggregate results is a good idea, or is morally obligatory, does not follow from a demonstration that there are certain negative duties not to coerce or deceive participants. Shalowitz and Miller go on to claim: "It would be disrespectful to treat research volunteers as conduits for generating scientific data without giving due consideration to their interests in receiving information about themselves derived from their participation in research". Giving "due consideration" to their interests in itself does not of course mean that these interests must in all cases be satisfied, because an all-things-considered judgement may work against this conclusion. Crucially too, there are quite different accounts available of what really is in a person's interests. Moreover, we can see in this quotation a tacit reference to the claim that treating research participants as means to an end violates the notion of respect for persons – but a claim that does not stand up to full scrutiny. To refer to treating a person as a "conduit" for generating scientific data is to use language that conjures up images of merely "using" people; however, recall that no notion of respect for persons is against using people as means to ends, only using people *merely* as means to an end.

Shalowitz and Miller add another consideration, the idea that the information derived from research that is "about" the volunteers themselves is especially pertinent to the notion of respectful treatment, but they do not offer any substantiated account of why this might be so. This is a great pity, because it is here that a fleshed-out argument would be of great interest. The idea that information belongs to the individual from whom it derives, or whom it is about, is seductive but ultimately difficult to substantiate. The idea that genetic information in particular belongs to the person from whom it derives likewise has difficulties, including the difficulty of beckoning to a dubious genetic essentialism and the problems that we share genetic information with so many others.

Respect for Persons and Personal Information

There is a very alluring argument that information "about" an individual really "belongs" to that individual. This alluring thought doubtless underlies some of the debate about feedback of results to participants in research and in particular, the link that is so often made between feedback and an idea of respect for persons. Respect for persons, in its various manifestations, has as a core idea the notion that respect belongs *inherently* to persons, and so derives from an account of *what it is to be a person* (or on some versions, what it is to have humanity). Spelling out what it is to treat a person with dignity revolves around a notion of what a person, *qua* person, deserves; this may be elaborated with an account of what it is to be a person. In addition, on some accounts, what it is to show respect for persons is grounded in specific, concrete appreciations of the individuality of particular people (rather than being based upon general qualities that all persons share). Notions of identity – either of an individual meriting equal respect simply because of their nature as a person or of an individual with a specific identity – are close to these various notions of respect. Hence, for some it is a short step to claims that the information that grounds that individuality is crucial to maintaining that respect and additionally that the person may have an especial claim to control any such information. These claims are however problematic.

It is commonly held that scientific knowledge should be openly accessible to all. We tend to think about personal information, however, somewhat differently. Not only do we often think this is private, and that the individual should have rights of control over who should have access to personal information, that is, to prevent access, but also it is sometimes held that individuals have a positive right to access to personal information about themselves; or that it is disrespectful to withhold such personal information, for a researcher to have access to it whilst the individual whom it is "about" does not have this access.

The issue of access to information is discussed in more detail in Chap. 10 (See Sect. 2.2.4).

Ravitsky and Wilfond (2006) cites Partridge, Shalowitz, and Fernandez as grounding disclosure of results on "respect for participants". So, in citing past work, the argumentative momentum towards the unqualified claim that respect for persons does indeed ground return of results builds force. But, as we have seen, there are areas of great ambiguity and room for dispute in these arguments. It is also interesting that Ravistky cites here the rather general notion of "respect for participants", which alludes to but is not the same as a philosophically grounded notion of respect for persons. A general notion of "respect" is hard to gainsay – for who would be against it? But it is philosophically and ethically problematic to use this as a generalised and inchoate ground for unspecified action; without further detailed argument it is not a robust basis for substantive claims.

Knoppers et al. provide an extensive analysis of international guidelines on the return of results to research participants and on this basis claim that there is an "emerging ethical duty to return the results of research" (2006). They have, in fact, only shown that there is the emergence of statements of such a duty. This does not in itself mean that this alleged duty is obtained or based upon sound reasoning; as we have seen, many of the arguments purporting to ground such a duty are based upon highly problematic or thinly sketched accounts of respect for persons. We have to be careful that we are not faced with argument by numbers; we need good arguments, not just several of them.

Knoppers et al. themselves recognise that there is a great deal of confusion in the claims. "Guidelines are neither standards nor laws. But, if over time they become professional norms, they may become the legal standard of care for both clinical practice and for research". But this is in fact a statement of the problem! They may become the recognised standard of care by default even though they are poorly grounded.

However, Knoppers et al. claim: "The ethical principles of respect for the person, beneficence and justice obligate the researcher to offer results in a manner that is clear and understandable to the research participants" and this is stated as a fact. In support of this claim they cite Fernandez et al. for this – which as we have seen, provides only a fragile basis for this claim. They also cite a European Commission report giving recommendations on genetic testing (European Commission 2004). But an examination of this document shows that it does not even contain any reference to respect for persons, nor does it deal explicitly with the return of research results. It is important to note too that Knoppers et al base their claim upon a conjunction of ethical principles: respect for persons, beneficence, and justice. It is not clearly stated how these principles might work together in such a case.

Respect for Persons, Beneficence and Justice

Respect for persons is often used, as is the notion of respect for autonomy, to signify the ethical mandate to take account of the value of individual persons.

Beneficence is the principle of aiming to bring about benefit: to address the consequences of actions. This is often taken along with the principle of nonmaleficence, summed up in the maxim "First do no harm". Aiming to maximise benefits may not be compatible with minimising harm, let alone with avoiding harm altogether.

Justice aims at some notion of desert, or of fairness; this may relate to different fields, including the distribution of goods, and may act as a limit on consequentialist approaches. The distribution of harms and benefits which brings about the best overall consequences may not be the fairest.

It is important to take careful note of the rhetoric of argument: how we are persuaded to certain opinions by use of language. It is easy to find in papers on this topic wording that implies there actually is a duty: "International bodies also recognise the ethical duty to disclose individual research results" (Van Ness 2008). To "recognise a duty" implies that there is such a duty, even that this alleged duty is well founded. But as we have seen, a well-argued foundation is hard to find.

There are also authors who use wording that implies the arguments are well founded. McGuire et al. claim: "Research ethics norms remain in constant flux. Nevertheless, convincing arguments have been made for allowing research participants to be permitted access to their personal data if they so choose" (2008). In claiming this, they cite Knoppers et al. (2006) – who as we have seen in turn base their claims on citing the weakly argued claims of Fernandez et al. (2003), as well as the collation of a number of varying guidelines. To give them their credit, the Knoppers et al. paper does give recognition of the complexity of the arguments and the different positions, yet in the end is cited simplistically as supporting the return of results.

This is a major methodological lesson of this chapter then, and one that can be generalised to many other instances. Respect for persons is an intuitively appealing notion; but the very fact that we have a rough and ready common-sense notion of what it involves can provide a danger, because this can lead us in argument to overlook its great philosophical complexity and the contested nature of different understandings of it. Without such careful scrutiny, it is a highly ambiguous concept that cannot with any robustness be utilised to ground detailed practical conclusions, tempting as this may be.

Moreover, there is a lesson to be learnt about citation of previous papers and arguments: be very wary of statements of ethical conclusions which are based upon summaries of other papers. The citation of papers towards a certain conclusion can otherwise produce a phenomenon akin to rumour, where a tower is built on very uncertain foundations. It is fascinating to see how the notion of respect for persons is treated as if it were simply an unproblematic and unified concept, whereas in fact this is far from true.

8.6 Concluding Unscientific Postscript: Problems with the Use of Science in Ethical Debate

I have looked at how an ethical concept, respect for persons, can be used and abused in debate in the ethical regulation of science. But good discussion in ethics also relies upon good use of science. It is importantly the case that sometimes just as ethical concepts can be taken out of context and without adequate characterisation or appreciation of their subtlety, so too can this be said of how scientific findings are used in ethical discussion.

When we are thinking about what developments might occur in the future, of necessity we need to extrapolate from the present and to think of hypothetical situations. These may be difficult to predict with much certainty. Scientific possibilities may be limited by a whole host of reasons including legal, regulatory, practical, and economic restrictions. The use of hypothetical examples can itself be instrumental in blurring the boundaries between the possible and the actual in genomics and acting to create hype. Especially where future possibilities are dangled in front of us gleaming with the allure of the ubiquitous future promise of genetics and genomics, this can in fact have the danger of skewing the debates: we think that what is around the corner is so marvellous, that no one must be denied it. But reality is often more mundane than our hopes.

Particular danger arises perhaps when what is claimed is not actually false. Let us look at an example:

"(1) New studies demonstrating correlations between common diseases and specific sites of genetic variation are appearing with ever-increasing frequency. (2) It is already possible, for example, to ask whether an individual has genetic variants associated with an increased risk of age-related macular degeneration, diabetes, Crohn's disease, cardiovascular disease, schizophrenia, and bipolar disorder, to name just a few. (3) Thus, every time one of these genome-wide association studies is conducted, the researcher theoretically has the opportunity to look in each individual's DNA not only for SNPs that correlate with the disorder in which she is interested, but also for any other SNPs that other investigators have identified as correlated with other disorders. ... [as well as] insertions, deletions, copy number variations all of which affect more than one base pair, and epigenetic modifications ..." (Clayton 2008, p. 288, sentence numbering added.)

This passage contains little that is false, but much that is potentially misleading. The first sentence is certainly correct, as studies showing correlations between common diseases have in fact been appearing at increasing rates, partly as a result of falling costs and scientific collaborations. However, note the reference not just speed of scientific progress, but of *increasing* speed. This frames the passage in terms of rapidly impending, significant progress. Note also the word "correlation"; many of the correlations found to date have not yet been demonstrated to be causally implicated in the conditions in question, and even if they are causal connections, the biological mechanism often remains to be discovered.

In the second sentence the opening gambit of increasing speed is here immediately followed by talk of the significance for individuals. A great many of the findings so far that associate genetic variants with common diseases, even where of serious scientific significance for understanding the underlying disease biology, lack any significant relevance for individuals (Heard et al. 2010; Frueh et al. 2011). The second sentence hence is not false; it is merely highly misleading. The third sentence however is partly misleading, and partly false. It suggests that in any study there is the theoretical opportunity to find all other variants identified so far as correlated with disease. But this is simply false in many studies, unless the entire genome had been sequenced. Moreover, there are a host of practical difficulties in looking for every known correlation in every research project. Quite simply, research would grind if not to a halt at least to a crawl were this to be done.

Let us now turn to another example:

"Genomic technologies are increasing the range of genome scrutiny. Millions of data points can be collected on every individual. Whole genome sequencing (6 billion base pairs) is not far behind. Once the entire sequence of an individual is known, every genetic variation will be revealed. Finding the useful information will involve identifying variations highly associated with disease or significant health issues" (Van Ness 2008) p. 297.

This is a good example of a mixture of enthusiasm about the advances of science and technology that then is followed by a misleading statement about its power. "Scrutiny" is a word suggestive of closeness and rigour but in this context scientifically imprecise. To state that "once the entire sequence of an individual is known, every genetic variation will be *revealed*" is again highly misleading. In one sense, sequencing the whole genome will give "all" the genetic variations. But the notion of "revealing" information suggests that it is fully understood. Finding DNA at a murder scene does not in itself "reveal" the killer. A moment's reflection will expose the absurdity of this notion: it suggests all the science has already been done, but we are talking about DNA collected for research, because much of the science is yet to be completed. Arguably, such writings implicitly trade on an imagined future where far greater power is achieved than is possible currently or in the near future.

Passages such as those we have studied have the general effect of magnifying the possibility of making such discoveries in a way that reinforces the notion of genomics as extremely powerful – and as extremely close to clinical application – when actually we are discovering fast also how much we do not really understand – and when much current genomics research is far removed from future application. There is an interesting subtext: much of the literature also reinforces a misleading picture of the scientist somehow being unimaginably close to vast amounts of information that is "about" another person – producing a metaphorical, imagined picture of power and control over another individual's destiny. But, in many ways the way the science is conducted is the reverse of this. Scientists often work in large networks, and usually with aggregated and anonymised data often collected at a different site and sometimes long ago.

What implications might be drawn from these accounts of the science? Often, this has the effect of being used to argue for an increase in the responsibilities of researchers (which will be the topic of the next chapter).

Bovenberg et al. state "... the ability of companies like deCode to translate scientific findings into personal health information for their customers, in a matter of weeks, does raise the question of why population biobanks decline any translation of their findings into personal health information for their participants" (2009) p. 56.

This is an extremely interesting passage which raises the important question of how changing social circumstances, outside the direct control of the research community and of individual scientists, may be seen to change expectations in ways that then may create implications for ethical practice. The rhetorical question is raised by Bovenberg, dangling in front of us the suggestion that biobanks are being remiss. But a detailed answer to the rhetorical question can be given. Direct-to-consumer (DTC) companies such as deCodeMe precisely give results to paying customers because that is their business and main aim; this is not the aim of scientific research. Furthermore, the simple description of translating "scientific findings" into health "information" papers over a multitude of questions. There are major doubts held by many over the real significance of many of these results for the health of individuals (Frueh et al. 2011). To equate such genetic results simply with "health" information is to ride roughshod over this legitimate ongoing debate of the relevance of much DTC data to health.

The language chosen is also telling: to describe population biobanks as "declining" to give findings to individual participants is to suggest a stubborn refusal that might not be suggested by use of more neutral language such as "do not provide". Indeed, it is also worth pointing out that one could equally ask this question in reverse: what does the policy of population biobanks not to return findings to participants say about the practices of DTC companies? Many voices are saying that currently, such companies are selling much information that is of scant value.

Shortly following the passage quoted above, Wright Clayton makes the claim: "Investigators' decisions not to obtain the software needed to examine individual results beyond those under study and manufacturers' choices not to put on chips disease-associated SNPs with those needed for the research are not likely to be viewed with favour by those research participants who want to receive incidental findings" (p. 288).

Careful attention to language is telling. The notion that investigators "*decide not to obtain* the software needed to examine individual results" suggests an act of deliberation over whether or not to do this. The writers may not have intended this, but it almost sounds as if investigators are purposefully flying in the face of research participants' wishes and interests. Likewise, it sounds as if manufacturers are positively choosing to leave off disease-associated SNPs. Consideration of which SNPs to include is more likely to be guided by a mix of scientific considerations and practicalities; sometimes chips are custom designed for particular types of research and sometimes generic ones are used.

It is also of interest to note how the example of BRCA1 is used. This is an oft-chosen example of the issue of feedback. For example, it is used also here, in a short article in a genomics ELSI newsletter:

"Alice signed up as a 'healthy control' for a research project into the genetics of type 2 diabetes. During the project, researchers identified a variation in Alice's BRCA1 gene that is known to be associated with a high risk of breast cancer. Alice is unaware that she carries this variant, and if she was told about it she would be able to take steps to minimise her risk of cancer" (MacArthur 2009).

As I argued in Chap. 1, choice of disease exemplar can skew any general conclusions drawn. There are many relevant differences between different genetic diseases that potentially affect responses towards them. They vary in terms of severity; age of onset; symptom pattern; transmission pattern; treatment availability, acceptability, and cost; stigma; and emotional factors. It is interesting that various forms of cancer are often cited in illustration of the issue of feedback of results, and breast cancer in particular. Cancer is an especially emotive disease, and breast cancer perhaps especially so. As Susan Sontag has shown in her work on illness as metaphor, different diseases can be subject to powerful undercurrents of thought that can radically affect how we think and feel about them; cancer is repeatedly thought of in terms of warfare, of battle, of struggle (1983). Those who die with this disease are routinely described as having died after a "long battle". Hence, we proceed into the debate about feedback of results against the unspoken background assumption that all weapons in the armoury against cancer must be drawn into play. Whilst we are unlikely to escape these metaphors entirely, it is worth bearing this in mind in choosing and responding to examples to illustrate issues. Such brute responses can easily distract us to underlying practical, economic, scientific, and technical considerations.

8.7 The Need for Detailed Consideration of Different Cases

I hope to have shown that it cannot simply be argued that accepting some notion of respect for persons *in itself* justifies a policy of feedback of the results of genomics research. This is not so far to have answered the question about feedback, however. We need to look extremely carefully at a host of considerations and at how different situations differ when extrapolating from one to another. Arguments for feedback of individual results in one area of research may start from examining practices in other areas of research. Often research involving detailed body imaging, such as MRI scans of the brain, is used as a reference point. Such technology has the potential to reveal, for example, a previously unknown brain tumour; it has been quite reasonably held that under these circumstances, it is a moral imperative to feedback such a result to the research subject. The close proximity of the researcher to the subject and the seriousness of the findings are here both greatly relevant. Moreover, in such research it is routinely part of the consent process that recruits are carefully told of this possibility.

In other areas of research, we may be faced with situations which differ in many ways. Importantly, even within genomics research, there will be a multiplicity of different situations. A genomics project may involve close dealings with a small group of people recruited from one hospital or clinic or a large international collaboration using archived samples. Consent forms may not have mentioned the possibility of feedback – raising in some cases the question of whether a subject might want to know. In other cases, consent forms may have explicitly stated that no feedback would be given. Research findings in genomics may be of great potential significance for disease prevention and treatment or reproductive decisions, but they may be of uncertain significance, or there may be no known treatment or prevention strategies. They are almost certain to need verification, as standards of accuracy will not be the same as for clinical testing, with implications for costs as well as for the potential to create worries from large numbers of false-positive results. It is also not the case that researchers will necessarily be sifting through *all* the possible results for *all* the recruits; moreover, the significance of any results is going to be changing rapidly over time as our understanding of genomics increases. It has also been pointed out that there are high cost implications for many suggested policies of feedback, and that in effect we may be led simply to an extremely inefficient way of doing genomics testing on a sample of the population who volunteered for research and met the recruitment criteria – not necessarily good public health policy. These factors will vary with the project – its consent conditions, the time since recruitment, its financial resources – with the results found – the condition in question, the degree of risk, and the possibility of treatment or prevention – and with time – as our understanding of genomics increases, and also as changes develop in treatment possibilities.

8.8 Conclusions

In this chapter I wished to show:

How in reasoning in ethics it is important to be clear about the exact meaning of important concepts used, especially ones imbued with value and salience; how terms can be used in different ways; with vagueness; and with ambivalence and ambiguity;

How it is vital to follow back references and reconstruct how arguments are built upon the claims of others;

How science likewise can be used inappropriately, and in particular how misleading impressions can be created even though no untruths are uttered

How these faults can lead to an escalation of the hype that then in turn can drive the ethical juggernaut in certain directions;

How therefore we must conclude that debates about the feedback of results in research need to be carefully reconsidered.

An additional lesson to be taken from this is to understand how complex these questions can be and how embedded in particular historical, local and cultural contexts. An understanding of this complexity should raise questions about the certainty of other judgements we might be inclined to make.

We need to be mindful of the difficulty of such questions, and realise that in many situations, it is genuinely difficult to find the right answer.

8.9 Study Questions

1. A genomics research project uncovers findings of potential significance to an individual participant. For this exercise, suppose that the consent forms for this project had made no reference to the issue of feeding back findings.

   (a) What are the arguments *for* and *against* feeding back results in the following cases:

   A variant indicating a 10% increased risk of coronary heart disease compared to the population average.
   A variant indicating a 10% increased risk of multiple sclerosis compared to the reference population average of 1 in 800.
   The BRCA1 variant giving increased risk of breast cancer.
   A variant of the ApoE gene giving increased risk of Alzheimer's disease (see Sect. 2.7.3)
   The mutation for Huntington's disease.
   A balanced chromosomal translocation which might cause problems in the next generation, to a woman in her early fifties.
   For a study where families are recruited, results indicating misattributed paternity.
   For a study where families are recruited, results strongly suggestive of possible incest.
   For each of these, you may need to do some research work to find out more about the conditions. Things to consider may include severity of disease, treatment options, and prevalence of disease.

   (b) What are your reasons for your answers? Do any of these reasons relate to the notion of respect for persons?

2. Think of ways in which you think people have been treated with disrespect. Try to work out what, if anything, is common amongst these cases. What would have to be different for the people involved to have been treated with respect?
3. What might it mean to treat someone only as a means? What might it mean to treat someone as an end in themselves?
4. Ways of showing respect vary greatly from place to place and time to time and between different people varying, for example, with social roles. Illustrate some of these ways;

then think about what a common core of respect that *all* human beings deserve might be.

5. You are on holiday on the beach when you pass someone sitting alone and notice that they have a very suspicious-looking mole on their back. Do you stop to warn them about this?

Think how your answer might change in the following cases:

A complete stranger
A complete stranger in a large family group
A complete stranger with two small children
A complete stranger who looks friendly
A complete stranger who looks annoyed
Someone you recognise from work but do not know personally
A colleague you know well
Your cousin
Your best friend
You are a health professional
You are not a health professional
You yourself have had skin cancer

Draw up a list of reasons for and against telling the person; list any reasons, not just "moral" ones – for example, "might save their life", "none of my business" "I might be wrong" "rude to interrupt", and "embarrassed".

8.10 Coming Up Next

In this chapter, we looked at how some notions of respect for persons, and arguments about the feedback of scientific findings, may act to increase the responsibilities of genomics researchers. The large networks of genomics research, and the often large distances in time and space between subject and researcher, make these questions pressing yet perplexing. We now turn to the issue of researcher responsibility.

Appendix: Participants at a Workshop on Respect for Persons, St. Hugh's College Oxford, 2010

Mhairi Aitken, University of Edinburgh
Louise Bezuidenhout, University of Exeter
Arndt Bialobrzeski, University of Marburg
Roger Bjugn
Meghan Bruce, University of Oxford

Isabelle Budin-Ljøsne, the Norwegian Institute of Public Health
Paul Burton, University of Leicester
Yu-mei Chang, University of Oxford
Alan Doyle, NIGB Director
Stephanie Dyke, Sanger Institute
Khaled El Emam, University of Ottawa
Claudia Emerson
Grethe Foss, Norwegian Biotech Advisory Board
Donna Gitter, City University of New York
Bryce Goodman, University of Oxford
Edward Gosden, University of Oxford
Jane Green, University of Oxford
Alison Hall, PHG Foundation
Shawn Harmon, University of Edinburgh
Naomi Hawkins, University of Exeter
Nils Hoppe, Leibniz Universität Hannover
Tim Hubbard, Sanger Institute
Jane Juniper, AstraZeneca
Dorcas Kamuya, Wellcome Kilifi, Kenya
Barbara Koenig, Mayo Clinic
Ayumi Koso, the Database Centre for Life Science
Jacob Leveridge, Wellcome Trust
Matteo Macilotti, University of Trento
Aisha Malik, University of Oxford
Kenji Matsui
Karen Melham, Oxford Bioethics Network
Mari T. Minowa, the Database Center for Life Science
Madeleine Murtagh, University of Leicester
Kieran O'Doherty, University of British Columbia
Saminda Pathmasiri, Public Population Project in Genomics
Cedric Pruski
Emmanuelle Rial-Sebbag, Genetics and Society Platform
Mark Robertson, AstraZeneca
Behnaz Schofield, Universal of Bristol
Nayha Sethi, University of Edinburgh and SHIP Project
Liliana Siede, Adviser National Congress of Argentina
Carol Smee, Sanger Institute
Liliana Soto Gomez, University of Leeds
Anne- Marie Tasse, Centre of Genomics and Policy, McGill Universit
Paulina Tindana, University of Oxford
Prodromos Tsiavos, LSE
Louise Vale, Consultant
Christine Wasunna, Kenya Medical Research Institute
Edgar Whitley, LSE
Don Willison, OAHPP

Stefan Winter, Hannover Medical School
Jessica Wright, Leicester University
Vicki Xafis, Public Health, University of Adelaide

References

Bovenberg J, Meulenkamp T, Gevers S (2009) Your biobank, your doctor? The right to full disclosure of population biobank findings. Genomics Soc Policy 5(1):55–79

Clayton EW (2008) Incidental findings in genetics research using archived data. J Law Med Ethics 36(2):286–291

Dillon RS (2010) Respect. Stanford encyclopedia of philosophy. Stanford University Press, Stanford

Downie RS, Telfer E (1969) Respect for persons. George Allen and Unwin Ltd., London

European Commission (2004) 25 Recommendations on the ethical, legal and social implications of genetic testing. Brussels

Fernandez CV, Kodish E, Weijer C (2003) Informing study participants of research results: an ethical imperative. IRB Ethics Hum Res 25(3):12–19

Frueh FW, Greely HT, Green RC, Hogarth S, Siegel S (2011) The future of direct-to-consumer clinical genetic tests. Nat Rev Genet 12(7):511–515

Heard E, Tishkoff S, Todd JA, Vidal M, Wagner GP, Wang J, Weigel D, Young R (2010) Ten years of genetics and genomics: what have we achieved and where are we heading? Nat Rev Genet 11(10):723–733

Human Genetics Commission (2002) Inside information: balancing interests in the use of personal genetic data. Human Genetics Commission, London

Kant I (1948) Groundwork of the metaphysic of morals. Hutchinson University Library, London

Knoppers BM, Joly Y, Simard J, Durocher F (2006) The emergence of an ethical duty to disclose genetic research results: international perspectives. Eur J Hum Genet 14(11):1170–1178

MacArthur D (2009) It's my genome: should researchers be obliged to return genetic data to research participants? Robinson Bradshaw and Hinson. http://www.genomicslawreport.com/. Accessed July 2011

McGuire AL, Caulfield T, Cho MK (2008) Research ethics and the challenge of whole-genome sequencing. Nat Rev Genet 9(2):152–156

Partridge A, Winer E (2002) Informing clinical trial participants about study results. J Am Med Assoc 288:363–365

Ravitsky V, Wilfond BS (2006) Disclosing individual genetic results to research participants. Am J Bioeth 6(6):8–17

Shalowitz DI, Miller FG (2005) Disclosing individual results of clinical research: implications of respect for participants. J Am Med Assoc 294(6):737–740. doi:10.1001/jama.294.6.737

Sontag S (1983) Illness as metaphor. Penguin Books, London

Van Ness B (2008) Genomic research and incidental findings. J Law Med Ethics 36(2):292–212

Chapter 9
Responsibilities of Researchers in Genetics

The duty and responsibility for ascertaining the quality of the consent rests upon each individual who initiates, directs, or engages in the experiment. It is a personal duty and responsibility which may not be delegated to another with impunity.
Nuremberg Code (Trials of War Criminals before the Nuremberg Military Tribunals under Control Council Law No. 10, Vol. 2 1949).

Abstract What responsibilities do those engaged in genomics research have? It is vital to address this question, but the answer is unclear. In this chapter researcher responsibility is explored by examining responsibility in health, through comparison and contrast with the responsibilities of medical practitioners, with a brief consideration of the responsibilities in law, and philosophical consideration of the nature of responsibility. There have been recent calls for researchers in genomics to take on expanding responsibilities, but there are genuine difficulties, created for example by the large size of many research networks. Researchers may have specific responsibilities not just to particular research subjects, but also to the wider community. The responsibilities of research subjects are also considered. Just as we have seen in the case of individual autonomy, the wider context of responsibility must be addressed. Some key research findings from moral and social psychology about taking on or avoiding responsibility are considered.

9.1 Introduction

Amongst the many questions in ethics and regulation posed by developments in genomics research, there are many pressing questions about the responsibilities of researchers towards participants in research and elsewhere. These questions are becoming increasingly important for many reasons. Developments such as widespread data sharing, long-term storage of data and samples, and international collaboration in studies with very large numbers of subjects spread across different countries and continents all greatly increase the numbers of individual participants

P. Boddington, *Ethical Challenges in Genomics Research*,
DOI 10.1007/978-3-642-23699-0_9, © Springer-Verlag Berlin Heidelberg 2012

who are involved in various ways with genomics research. Furthermore, genomics potentially raises issues not just for the participant but for family members (Cassa et al. 2008; Botkin 2001) and population groups (HapMap Consortium 2004). The reach of genomics research extends beyond medicine to non-medical phenotypic information and potentially raises issues around ancestry and relatedness as well as having potential forensic application. Moreover, such research is usually carried out in large and interconnected groups and individuals within these groups may have responsibilities not just to research subjects but to fellow workers. Such developments together expand the potential sphere of these as yet uncertain responsibilities.

Like calls to respect persons, calls to take on responsibility are seductive, for who could be against being responsible, who would want to be thought "irresponsible"? Yet there is ample scope for different understandings of what responsibility involves and of whether particular people have particular responsibilities in certain instances (Glover 1970). "Responsibility" is one of those central ethical concepts often used with deleterious or even disastrous looseness. We have also looked closely at problems with the operation of ethics within a social and institutional context, and at how easily we may diffuse responsibility away to others (Chap. 7).

A laudable and appropriate call is made for individual researchers to take on a personal responsibility for the ethical conduct of research. However, the very formality of ethics regulation, coupled with the large size of many research groups, means that there has developed in many cases a potential schism between the consciences of the individual researchers and the formalised, "external" demands of ethics regulation. This presents a genuine contemporary dilemma which it is not easy to circumnavigate.

It is vital that any apportioning of responsibility is both fair and reasonable. If researchers in genomics are to be called on to exercise responsibility in various ways, then the nature and extent of their responsibilities must be delineated as clearly as possible. Especially where practice is developing and where science and technology are progressing, there is always going to be a place for a range of views and for interpretation about what responsibilities individuals, groups, and institutions have. Moreover, responsibility without the corresponding capacity to act is both empty and potentially dangerous. For example, some of the ethical issues touching genomics research arise from a wide context of ethical and social issues, such as global inequalities, about which genomics researchers *per se* have little control (Chokshi and Kwiatkowski 2005). Therefore, it is important to examine calls to responsibility rigorously.

We have looked at how researchers operate within institutions and within a larger context of regulation and national and international law. Researchers potentially have multiple moral demands. Some of these stem from the requirements of ethics regulation, some will stem from the requirements of science, but others will stem from a researcher's own conscience, moral beliefs, and outside commitments. This can further complicate judgements of moral responsibility in research.

In this chapter we will look at the question of the responsibilities in genomics research by careful analysis of the notion of responsibility. We will be concerned with the nature and extent of responsibilities within genomics research, and in looking at this, will examine the problems and challenges of encouraging and allowing morally responsible behaviour, both generally and within the specific context of the workplace especially of course within scientific research work. A discussion of the ubiquitous and varied notions of responsibility in health forms a background that can shape how the particular question of responsibility in genomics research is debated. The responsibilities of doctors are far more clearly specified than those of researchers, so will be a useful starting point of comparison and contrast. Philosophical, legal, and psychological notions of responsibility will then be drawn upon critically to examine current calls for genomics researchers to take on responsibilities towards participants. It is also necessary to look at the corresponding question of the responsibilities of participants.

9.2 Responsibility for Health

Public discussion of health is rarely far removed from implicit or explicit assumptions about responsibility. These are often misplaced, and moreover often swiftly followed by accusations of blame, whether spoken or implied (Boddington 2009). Heated discussion of responsibility can be seen in debates such as those concerning individual "life-styles" that lead to increased risk of various diseases; of governmental responsibility for taking action on many of the causes of ill-health, such as environmental pollution, social exclusion, and childhood poverty; debates about health and safety regulation, and bans or restrictions on the marketing and labelling of products such as fatty foods, alcohol, and tobacco; the responsibility of parents towards their children; attitudes towards screening for disease including pre-natal screening; debate about responsibility for the harms of addiction; in discussions over socialised medicine such as that occurring in the USA and that occurring over incursions into socialised medicine in countries where it currently exists; and of course in the frequent accusatory cry that we are living in a "Nanny State" (Sachs 1996). From this brief survey, the range of individuals and organisations alleged to have certain responsibilities in relation to health is wide. It includes governments, both national and local, including areas of government such as transport, agriculture, environment, and education; industrial producers; parents, including the pregnant; employers; the sick; the well. Responsibility may be assigned for causing, preventing, curing, worsening, or neglecting ill health; for health, and for disease; and, in case that list was not long enough, for being overly concerned with health (Fitzpatrick 2001).

Whilst our discussion cannot look in detail at global issues about responsibility for health, it is worth bearing these in mind, especially as these can form an unconscious frame to the issues under consideration. It is a background question that shapes the general issue of what responsibilities researchers in genomics might owe to participants of research or to society more generally, and to the question of any responsibilities that members of society may have to participate in medical research.

Philosophical Perspectives on Responsibility and Blame

Philosophers have spilled much ink and possibly many tears on debating the nature of responsibility. The word "responsibility" is one where we are often bewitched by careless use of our language into drawing unwarranted conclusions.

It is helpful to distinguish **causal responsibility** from **moral responsibility**. The former simply means that a person, or indeed an inanimate object, animal, or other natural phenomenon, has had a causal role in bringing about a situation. Moral responsibility implies some judgement or obligation: but it is important to be clear that much will depend on the particular circumstances. To have moral responsibility may mean that one may have responsibility to address a situation, that is, to right a wrong; one may be praised or blamed for a situation; one may be held to account by others for one's actions (Glover 1970).

It is important to be clear that causal responsibility does not imply moral responsibility, nor does moral responsibility necessarily imply causal responsibility, although these are often closely linked. One may have caused a bad situation but may be excused, for instance, if one acted out of excusable ignorance or from genuine loss of control, being pushed by an external force, for example. And one may have moral responsibility for addressing a situation one did not directly cause, for instance, if one allowed it to occur through negligence or neglect of a positive duty; or if one's role means that responsibility is assigned (for example, it may be your responsibility to take out everybody's rubbish).

However one of the most vexing questions in philosophy – and in other areas such as law and psychiatry – is the question of when a person is or is not "really" responsible for their actions. Cases where this may not be clear include some cases of mental illness or personality disorder, and "borderline" cases such as the question of at what age a child can be said to be fully morally responsible for their actions.

One of the most important things to bear in mind is how swiftly responsibility is linked to often rather crude notions of blame. For instance, discussion of the causal role of nutrients and other substances on foetal growth is routinely met by heated debates and accusations that this amounts to "blaming the mother" (Warin et al. 2011). But even if one is held morally responsible for a bad situation, this may mean that one is in some way held to account – that one is accountable – and may be called upon to address the situation in some way – yet this need not be in any punitive or blaming way. Understanding and tolerance or forgiveness can in some instances go hand in hand with attributions of moral responsibility.

The culture of blame can be problematic in institutions where it is important that mistakes are uncovered. Steps have been taken in recent years within the NHS to instil a culture of openness about errors, removed from the stigma of blame in order to foster responsibility without blame, to increase the reporting of errors, and to learn more effectively from any errors (Donaldson 2003). Whilst there are questions about how successful this has been, the need for such a policy demonstrates well the quick step often taken to blame, and some deleterious effects.

Responsibility can be a two-edged sword. To take on responsibility can be empowering and humanising (Sartre 1973). It may spur us to great things, and may make us feel like valued and active members of a community. But to attribute responsibility to someone with limited control over a situation can be demeaning and even dehumanising – and work in psychology backs this up as we shall see. Responsibility without adequate control is now widely recognised as a major cause of workplace stress (Stansfeld and Marmot 2002).

9.3 Responsibilities of the Medical Profession

The duties of a doctor towards their patient are of course themselves subject to debate and development, but in comparison to those of the medical researcher, they are relatively clear and delineated, not just in the ethical regulation of doctors but also in law (Mason et al. 2002). Regulations differ slightly between countries, but there is broad agreement on certain major points. It is quite clear that doctors have responsibilities first and foremost to patients in their care to provide adequate and appropriate medical treatment, and to respect the relationship between doctor and patient. This means that patients can rely on assurances of confidentiality of medical matters, and that doctors will provide patients with information so they are able to give adequately informed consent to treatment, and that they will practice according to acceptable professional standards and refer patients to other doctors and for treatment as and when necessary (General Medical Council 2006). In law, such duties are laid down in terms of a "duty of care" between practising physicians and their patients.

These responsibilities place the health of the individual patient at the pinnacle of concerns: what is at issue is the relationship between an individual patient and a doctor (although in practice this extends to a wider health care team). However, there are certain limits to this, but these are relatively clearly defined. For instance, the duty of confidentiality may on rare occasions be overridden when this is necessary to prevent a clear danger to another, for example where child abuse is suspected or where there is a clear and imminent threat of harm to another. The requirements of public health may also conflict with and sometimes override the rights of individual patients, such as when enforced quarantine or treatment might be deemed necessary during an epidemic.

Genetics has posed difficult problems for defining the responsibilities of doctors because of the potential for genetic information to have implications for genetically related others – who may or may not be patients of the same doctor. Medical professionals may often be in the position of suggesting to patients that they should pass on genetic information to others. However, patients sometimes decline to do so. Guidance in recent years suggests that there may be a moral obligation on patients to do this and that doctors may be advised to try to persuade patients to do this (General Medical Council 2009).

Interestingly, research shows that many patients tell medical staff they are going to pass on this information to family members but do no such thing. But this in itself is not a failure of responsibility but may be based upon a careful consideration of wishing to protect others or of judging when the right time to tell is (Arribas-Ayllon et al. 2011). This is a very good example of how someone may be seen to be "irresponsible" in not passing on genetic information, when in truth their actions are based upon a well-considered but different notion of what it is to be responsible, given their more detailed knowledge of their family circumstances.

It should also worth be noted that the duty to provide an appropriate standard of medical care has its limits: manifest in economic limits placed on treatment by health insurance companies or by the National Institute for Clinical Excellence's (NICE) National Health Service guidance on spending on treatment, and manifest too in waiting lists for treatment. Since it is the case that the world's resources are finite, such limits are not in themselves morally inappropriate, although on occasion they may be.

9.4 Responsibilities of Medical Researchers

The responsibilities of medical researchers are complex and may pull in different directions. They have responsibilities towards research subjects which bear some parallels to the responsibilities owed by doctors to their patients, but which are much less clearly defined. They also have responsibilities concerning the research enterprise itself. Intellectual responsibilities to conduct valid, robust research arise which may at times be in tension with other duties towards participants. These twin domains of responsibility demonstrate how researchers on human subjects are heirs to the values of the Enlightenment. The pursuit of knowledge, and the release of such knowledge into the public domain for the perusal, use and benefit of all, is a key value of the Enlightenment. The protection of the life and liberty of the individual is another key Enlightenment value; at times, however, these may come into conflict.

Duties of confidentiality and of informed consent to research parallel those of doctors to their patients, with some variations in how these are implemented and understood. In clinical medical research, patients are being given experimental treatment. In these cases, standards of informed consent are high, and duties of care to the research participant as patient are still in place. For instance, should it become apparent that the experimental treatment has seriously deleterious effects, or conversely that it is extremely effective so that those receiving only a placebo may be losing out on significant health gains, research may be halted early. Nonetheless, in such clinical research, the gains of research are balanced alongside duties of care to patients. It is significant that there may be some conflict. Finishing a clinical trial early may benefit those who would otherwise only get a placebo, but it may pose a threat to the robustness of the conclusions of the trial. There is, in at least some situations, an inescapable conflict between the responsibilities to conduct good scientific research in a clinical context, and responsibilities towards patients.

However, in other medical research, participants undergo no intervention. Such research includes epidemiological research or other purely observational research, and includes much genomics research, where the collection of a small blood sample constitutes the only real medical risk. Hence there are substantial differences in the responsibilities of the researchers. Duties of confidentiality still arise, but the relationship is arguably rather unlike that of a doctor to his or her patient. There is no intention to treat or to offer treatment or advice for a health condition.

Indeed, it may be thought advisable clearly to differentiate the role of the researcher from the role of the clinician in order to avoid the "therapeutic misconception": the common perception of those enrolled in research that they stand to benefit personally from participation (Appelbaum et al. 1987). It has often been found that this illusion has a tendency to persist even despite statements to the contrary. It might be said then, that the primary duty of the researcher is to do research, and whilst they may have responsibilities to avoid various types of harm to participants, they do not have any therapeutic responsibilities to treat. The lack of clarity of the roles and responsibilities of researchers with regard to therapy is in fact a central issue in the current debate over feedback of results to research subjects discussed previously.

The responsibilities one human being has to another are shaped in large part by the nature of the relationship between them. In many forms of research, the relationship between researcher and participant is quite unlike that of the traditional personal and individual doctor-patient relationship. There may be significant differences between different research projects.

Relationships between subjects and researchers may vary. There may be variations in how people are recruited to a project. In longitudinal studies and some cohort studies, recruits may have a long-lasting relationship with a group at a particular institution who initially recruited them and who maintain continuing research relationships. In some instances, this may involve elements of treatment. In other studies, collection of samples and data may take place at one point in time and there may be no actively continuing relationship.

There are variations in the staffing structure of projects. In many projects, although there are principle investigators who take overall responsibility for the research, there are some who are involved in recruitment and others involved in other aspects of the research. There are often networks of researchers within larger networks. The researcher may never even have met the participant: the identities of participants are generally kept confidential via a system of anonymisation; indeed very many researchers may have access only to aggregated data and not to individual data at all. These features of research may have relevance in ascertaining the duties and responsibilities of those involved. Institutions within which research is conducted may have particular responsibilities some of which are clearly codified.

Boundaries between research and treatment may vary. There may be variations between "pure" research and research involving links to treatment (a boundary potentially blurred by the feedback of results). Genomics research may take place as part of a larger research project, involving different elements including treatment, but it often involves no clinical element and it frequently occurs in large consortia where there are many researchers most of whom will have had nothing to do with the recruitment of participants.

The numbers involved may vary. Research in genomics, like much epidemiology, may involve extremely large groups of participants. And, especially given that any risks of genomics research are related to the information generated, this information is potentially of relevance to much larger groups – to genetic relatives and also to genetically related population groups. Genomics information is not unique in this regard – information that an individual is the carrier of an infectious disease has implications for others. But the concrete nature of the overlaps in DNA between related individuals, and the power of some techniques to link family and population groups through mapping DNA, provide particularly vivid links of one individual to others.

The scope of any responsibilities of genomics researchers thus it is sometimes argued may extend very far. There are more researchers; there are more participants; and genomics science and technology have rendered possible very powerful uses of the information generated. In genomics research, we are faced with an exponential increase in the number of researchers, holding different positions within large and often international networks, and in the number of research participants, bringing into the picture genetically related others who may also have an interest in the generation and use of genomic information. Awareness of the implications of genomics research for these serried ranks, together with the potential power of genomics information, has led to suggestions in the expansion of the responsibilities of those involved in genomics research. These suggestions call for careful critical scrutiny.

9.5 Recent Calls for Researchers in Genomics to Take on Expanding Responsibilities

It is quite clear that the potential for the list of responsibilities of genomics researchers to grow is large, and these calls have been made in a wide range of contexts. Calls for the expansion of the responsibilities of genomics researchers have taken place in relation to individuals participating in such research; in relation to family members of those taking part; and in relation to communities and populations implicated in genomics research. These purport to expand both the extent and the type of responsibilities of genomics researchers. Each of these calls may be made with the earnest belief that expanding notions of responsibilities will raise ethical standards. But must it?

Calls to expand the nature and extent of responsibilities of genomics researchers with respect to populations included in research encompass, for example, issues concerning identity protection, concerning the naming of population groups; and concerning community resource allocation and rewards for research participation, suggesting that especially in relation to research in less developed countries researchers should take steps to ensure benefit sharing (Mascalzoni et al. 2008; HapMap Consortium 2004; Chokshi et al. 2006; Chokshi and Kwiatkowski 2005). Questions likewise have been raised about researchers' responsibilities towards the family members of those enrolling in research. It has been suggested that because the results of genomics research, and any informational risks thereof, may also have implications for related individuals, therefore families may need to be informed or consulted about participation in genomics research (McGuire et al. 2008). There is also a lively ongoing debate regarding responsibilities concerning the feedback of research results to participants in research, as we saw in the last chapter (Cho 2008; Caulfield et al. 2008a).

Benefit Sharing and Exploitation

Many call for the sharing of benefits of the fruits of research. These calls are especially vocal in the context of research conducted in the developing world. There are wide gulfs between the resources available to researchers in the developed world and colleagues in less wealthy nations and any recruits from the developing world. There are also concerns about the resources available to populations in general, especially medical resources, and concerns that developing countries may be less able to afford any medicines produced by research. There have been therefore worries about this great disparity and that the developed world may benefit disproportionately from the research. Consideration of benefit sharing hope to address this disparity (Schroeder 2007; Rotimi et al. 2007; Chokshi and Kwiatkowski 2005; Austin 2002; Knoppers 2000; Berg 2001).

This is also a good example of how social, political, and historical events and considerations from outside the immediate arena of research may be brought to bear upon a consideration of research ethics and responsibilities. The context of global inequalities and in particular, past histories of colonisation as well as contemporary political and economic relationships are an inescapable backdrop to these debates.

Moreover, a particular issue is that of the potential for genomics research to be cased in a racist light, as well as fears that the genetic information of particular groups is their rightful province to control.

(continued)

The claim is often made that exploitation may occur. This accusation may act as a powerful motivator to avoid any such charge. But what is it to "exploit" someone? As I have argued in Chap. 8, not all unequal treatment is *ipso facto* disrespectful. To exploit someone implies an intention to get more than you are entitled to, to deceive or mislead, or to abuse power. The backdrop of past history, and the continuing vast inequalities in the world, cannot be escaped but may make it hard to address the concrete question of just who has the responsibility for addressing past and present injustices throughout the world. What the responsibilities of individual researchers, research teams, research institutions, or funders might be to address these issues, and how far these responsibilities go, is a vexing question.

9.6 The Nature of Responsibility: "Chains of Responsibility" in Research Networks?

Caulfield et al raise the question of who might have a duty, whether moral or legal, to return results of research to participants, posing this question in the context of the different roles of researchers, from those who collect samples and data to those who produce results and who may not know the identity of participants. They ask "Is there is a 'chain of obligation' that runs through the various members of a research team?" (2008b). Cho likewise talks of a "chain of responsibility" for dealing with new findings of potential clinical significance; although she says that the extent and timescale of any such chain is unclear (2008).

The notion that there is a "chain of responsibility" raises extremely important issues about how responsibility might be distributed amongst a large and possibly loose network of researchers in genomics, and how that might extend to a large group of participants, family, and population groups, over a potentially long period of time. It brings to the fore issues discussed at various points throughout this book: from the embodiment of a direct responsibility between researcher and subject in the Nuremberg Code to the easy dissolution of question of responsibility discussed in Chaps. 6 and 7. With these difficulties in mind, any claims that there exists a chain of responsibility must be accompanied by some account of how such a chain would operate effectively within particular institutional and social contexts. Such accounts are as yet notable by their absence and would indeed be virtually impossible to deliver within an approach to ethical questions which confined itself to examining the standard questions of principles, codes, rules, and regulations. The social and psychological world within which such ethical notions operate is a necessary point of reference for such an investigation.

The Landscape Beyond: External Influences on Ideas of Responsibility
Our expectations are moulded by many things, including what is happening around us. Ideas about what is reasonable and what is unreasonable may spill over from one area to another. Conversely, it's often remarkable that we have great concerns in one area, but fail to generalise these concerns to other areas (for example, as we have seen, there may sometimes be inconsistencies in how worried people are about potential uses of their genomics information and how worried they are about potential uses of a great deal of other personal information).

What responsibilities we consider researchers in genomics to have may be shaped by outside developments and changing expectations. For example, it may be thought that there is a relatively clear line between research and clinical practice in many cases of genomics research, and that therefore the responsibilities of researchers are correspondingly different to those of medical practitioners. This may be used, for example, as an argument against feeding back individual results of research, on the grounds that researchers do not owe such a duty of care to research subjects as is owed by doctors to patients. However, there is a significant social development: it is now possible for people to access a great deal of information from direct–to consumer genetic testing companies. This outside development has led some people to ask: if consumers can get access to such detailed genomics information about themselves, why withhold it from research subjects? Some research projects may give feedback to subjects. The question then arises, why would other research projects not do the same? Ideas we have about the researchers' responsibilities to their recruits may be shaped and changed by these developments outside the control of the researchers themselves.

Likewise, especially in the context of research with populations from the developing world, or on indigenous populations or groups which have been disadvantaged historically, it may be impossible to ask questions about the responsibilities of researchers without considering the historical context and a wide range of social, cultural, and economic factors far beyond the direct control of the researchers themselves or of their institutions.

9.7 Responsibilities of Research Subjects and Genomics Research

There are quite different views on what responsibilities individuals may have to participate in genomics research. At one end of the spectrum are views that participation is entirely voluntary and even that those taking part merit rewards, such as payments or inducements to take part and a right to individual results. At the other end are views that individuals have a duty to take part in such research.

This may be explicitly linked to notions of civic duties, in particular that those who stand to benefit from health care should be prepared to contribute to the research enterprise that enables standards of health care to develop (Forsberg et al. 2009). This is a form of "fair exchange" argument, spread over time, for the anticipated benefits of current research participation are for the future, perhaps even for the fairly distant future, and the current state of medical expertise is built on research and practice some of which extends quite far back into the past.

It is manifestly the case that arguments about what responsibilities ordinary members of the public might have to contribute to medical research relate rather strongly to different health care systems in particular and to political systems of social welfare more generally. In countries with socialised medicine, where it is clear that medical research and treatment in a real way "belongs" to all, then notions of duties to take part in research may be inculcated more strongly than in those countries with private health insurance systems and privately funded medical training. This then is another instance demonstrating the need to look at wider systemic and social influences in addressing ethical questions. Without taking such important background factors into account, abstract debates about "duties" to take part in research may get nowhere.

We saw earlier on how our common human heritage has been appealed to by some in arguing for a notion of "genetic solidarity and altruism" grounding participation in research that could potentially benefit all (see Sect. 5.2.8). On the other hand, one may argue perhaps that appeals to notions such as genetic solidarity are metaphors calling for an imaginative ethical response to our shared human condition. There may be no special responsibility to take part in genomics research over and above any other responsibilities to participate in projects that contribute to the pot of common good out of which we all sup.

9.8 Legal Notions of Responsibility: The "Duty of Care" of Genomics Researchers

Legal responsibility is closely connected to moral responsibility although the two do not entirely overlap: moral responsibilities may extend beyond the law, and legal duties deal with formal or parochial matters or may be drawn up to standards of exactness needed for enforcement that may not have substantial moral basis. However, legal and moral responsibilities are sufficiently connected that a brief examination of the situation in law will help advance our discussion, as well as providing a useful test in cases of uncertain responsibility.

Legally, in the UK, there is no clearly established *duty of care* for researchers applicable to genomics (or other medical) research, whether for primary researchers or for secondary researchers who merely use data generated by others. Where there is no legally established duty of care, in UK common law the Caparo test applies:

Is the risk foreseeable?
Is there sufficient proximity between parties?
Is it fair, just and reasonable to impose a duty of care?
(Caparo Industries plc v Dickman (1990) 2AC 605).

The sense of this test should be clear, although of course there will be potential disputes about the exact scope of the terms. No one is expected to have any special powers to foretell the future, or, in UK law, to have to take on responsibilities for all and sundry. (There are great differences between legal jurisdictions on this important point; some legal regimes do in fact have "Good Samaritan" laws which require assistance to be provided to strangers in certain circumstances.) Using such a test could lead us towards a reasonable, proportionate, and workable notion of the responsibilities of genomics researchers towards participants. However, the possible areas of difficulty must not be underestimated. The notion of "sufficient proximity" between parties needs to be considered along with the difficulties mentioned earlier in discussing the notion of a "chain of obligation".

Article 2 of the European Convention on Human Rights (ECHR) protects the right to life, and it has been held that the UK is under an obligation to take "appropriate steps" to safeguard the lives of those in its jurisdiction. It has also been argued that Article 2 would place a positive obligation on member states to put in place measures to avoid a risk to life; and additionally that this could include the provision of feedback of risk of a serious genetic disease if this enabled treatment or prevention to occur (Johnson and Kaye 2004); this could apply to the researchers' institution or funder. There is still however confusion about this issue (Miller et al. 2008). One crucial question concerns the distinction between imposing a risk to someone's life, and taking steps to reduce risk to life. It is the latter which is in question in most of the difficulties envisaged regarding the extent of responsibilities of genomics researchers with regard to the right to life (see Sect. 4.5.1).

A duty of care on researchers to research participants has been established in litigation concerning research involving Creuzfeldt Jacob disease (Re Creutzfeldt-Jakob Disease litigation 2000). However, this case concerned a clinical trial: no cases as yet determine the scope of a duty of care in genomics research. A crucial element in determining proximity legally is whether a close relationship has been established and in law this relationship need not be very close or personal. Hence, it is possible that this may include those conducting research. It has been argued that in research such as longitudinal or cohort studies, such as biobanks, carried out over many years, then it would seem that such a duty of care does exist (Johnson and Kaye 2004) but this argument rests on conflating the difference between imposing a risk and warning of risk, and this point has yet to be tested in the courts. There are no decided cases of duty of care for statisticians and epidemiologists, who do not take on a clinical role. Many of the purported or disputed responsibilities of genomics researchers involve issues other than the right to life and there is little or no legal guidance here.

Slippery Slope Arguments

Slippery slope arguments – also often called "the thin end of the wedge" arguments – are common in ethics, although they frequently attract scathing sceptical criticism.

They take the following general form:

A proposed practice or act that opponents argue to be ethically blameless, or even ethically required, is characterised in a certain way;

It is argued that, once such a practice or act is allowed, there will be no way of preventing practices or actions that go further along some continuum: this is the alleged slippery slope.

At some point along the continuum, practices or acts acknowledged to be ethically wrong are reached.

There may be two sources of the alleged slide down the slope. In one type of slippery slope argument, there is held to be no **logical or conceptual line** that can be drawn to distinguish the initial step onto the slope from the wrong doing that lurks further down the slope;

In another type, it is held that it is empirically the case that such a slide will occur, because there is no firm **psychological line**, for example because moral scruples against a certain type of action will be blunted.

Those who argue against such warnings will then either argue that a logical or conceptual line can in fact be drawn to distinguish the permissible from impermissible and hence prevent the slide down the slope; or will argue that in fact, the empirical prediction of a psychological slide is mistaken.

Many routinely rubbish slippery slope arguments as inherently conservative scaremongering with scant basis in real behaviour. However, in the area of new developments in science and medicine there are many examples of slopes which have in fact been decisively slipped down. For example, when sex selection of embryos first became possible in about 1990, it was suggested that this was intended only to address X-linked genetic conditions, and it was mooted that "nothing would be easier" than to make rules to ensure this technology would never be used for "frivolous" selection of sex on social grounds or for parents' preference (Warnock 1990) which many took to be a misuse representing a slide down a slippery slope. Less than a generation later, the discussion of sex selection of embryos is still controversial, but is now a routine topic in bioethical discussion which many consider a valid option, and is in fact practiced extensively in some parts of the world.

Slippery slope arguments are highly relevant to questions of individual responsibility. Those who warn against such alleged slopes often do so precisely because of the danger that in travelling from one small point to the next along the slope it is easy to misrepresent to oneself the morally dubious points further down the slope and hence easy to evade responsibility for one's actions.

The reality and danger of slippery slope arguments is another theoretical and philosophical argument finding confirmation within empirical psychology. The ways in which justification for actions might proceed along such a slope finds strong echoes in much of the research into moral abuses discussed in Chap. 7.

9.9 Moral Responsibility: Some Lessons from Psychology

9.9.1 Warnings for Genomics Research: Ways in Which We Abnegate Responsibility

Under what circumstances will we take on responsibility? And when will we shirk responsibility? Here as in Chap. 7 we will look in some more detail at findings from moral and social psychology. Firstly, we will examine the danger that there may lurk in numbers.

Kitty Genovese is one of those tragic figures who have become famous only in death. She was stabbed in full view of many onlookers, gazing at the scene from their apartment windows, not one of whom reported the crime (reports of the exact circumstances vary slightly) (New York Times 1964). The stabbing was horribly drawn out, meaning that timely intervention may well have saved her. There has been much discussion of why exactly no one acted, but a generally agreed conclusion is in a situation calling for action, if there are plenty of other people around who could act, individuals are much less likely to act. We perhaps not unreasonably think: "surely somebody else will do it". But this can lead to nobody acting. The example of others failing to act can then reinforces this inaction – after all, nobody else is doing anything, so why should I? Empirical research and the example of other similar cases support this conclusion (Darley and Latane 1968; Moriarty 1975).

Given the often very large numbers of genomics researchers within a network, we need to give such findings careful consideration. Worries that individuals working in large groups may leave responsible action to others perhaps lie behind the calls for the creation of a "chain of obligation" in genomics research, whilst at the same time pinpointing the difficulties of achieving this.

We are also less likely to take on responsibility when we feel assured that another person, especially an authority figure, is accepting responsibility for our own actions. This was a clear finding of the Milgram experiments (Milgram 1974). Networks of research scientist of course do operate within hierarchical structures, with some having more authority than others, and overall authority being held by principal investigators, project leaders, and leaders of institutions and faculties. The willingness of the underling to hand over responsibility to authority figures however will mostly be a problem when the authority figure is malign – as in the Milgram experiment – or does not take their responsibility seriously.

The abnegation of responsibility on the part of authority figures can lead to what has become known as the "Night Shift" phenomenon (Zimbardo 2007). The worst abuses at Abu Ghraib, where Iraqi prisoners were tortured, subjected to sexual and other degrading abuses, and in some cases even murdered, occurred during the night shift. This was also found in the Stanford Prison Experiment, where the students acting as guards overnight were the most abusive. Why? There is a simple answer: because the authorities in charge were not around. At Abu Ghraib there was found to be a general neglect by authority figures to take control or even to become acquainted with the conditions on the ground, and far too few in charge of far too many.

In analysing what enabled subjects to deliver painful shocks to the "learner", Milgram argued that close attention to the component parts of action acted as a prop to moral disengagement; many subjects concentrated acutely on the mechanics of performing the task, directing attention away from what the task was actually accomplishing in shocking the "learner". This direction of attention to sub-units of a larger and morally problematic whole has also been remarked upon as an enabling device in other instances, for example, within Nazi Germany. The conduct of much research within large networks, and in particular the great divisions of labour that are seen in complex research, involving a number of highly specialised and technical roles, means that attention needs to be given to this issue of fragmentation and how it might impact upon ethical judgement and action.

We are also less likely to act on the calling of our consciences in situations where there is no "exit strategy" to escape from the situation that clashes with our conscience's calling. In the Milgram experiments, subjects had entered a contract to perform the experiment, and although in one sense they were physically free to walk out, in practice, to do so would involve explanation and argument with the experimenter, and the breaking of social expectations. Even these seemingly flimsy constraints can be enough successfully to encourage individuals to ignore their consciences, in other words, to act without responsibility. This should be serious food for thought in considering the constraints upon individuals within the workplace, especially in careers where individuals make large investments in their careers and where alternative employment opportunities may be scant.

A possible "exit strategy" could be the ability to speak up and be listened to when one's conscience is troubled. But one of the most alarming and instructive aspects of the Milgram experiments is the finding that subjects who spoke out to protest often nonetheless continued to deliver shocks. Milgram suggests that the very act of speaking out lessened tension, enabling subjects to think of themselves as good people, and actually therefore enabled the actions, about which the subjects had objected, to continue. On this analysis, vocal protests enabled bad action, rather than encouraging good action. Will the ability to speak out function as it did with many of Milgram's subjects, simply to relieve tension to the point that action discordant with beliefs is enabled?

When is responsibility too much? Key elements conducive to workplace stress include inadequate reward for disproportionate effort, and responsibility without control (Chandola et al. 2008; Stansfeld and Marmot 2002). Under stress, not only does the physical and psychological health of individuals suffer, they will also function less effectively. The role of responsibility without adequate control in producing the symptoms of stress provides one clear reason for careful attention to calls that researchers have certain responsibilities; if these are unrealistic, not only will such calls be ineffective, they may produce harm.

9.9.2 Ideology and Responsibility

Of the ingredients in the formula enabling poor moral behaviour, one bears particular reflection for anyone engaged in any kind of higher pursuit, as in medical research. The ingredient: an ideology that justifies behaviour. The end of the ideology can be then called upon to justify the means. Milgram's subjects carried on delivering shocks because they were acting in line with the ideology of the credibility and usefulness of academic research – even when without the trappings of a prestigious institution. We have also briefly considered how the ideology of health drives many (often competing) attributions of responsibility for health.

The problem is that sometimes it is entirely reasonable to say that the end justifies the means. We are used to this in medicine as something so routine it is hardly even noticeable. Cutting into another person's body is, in itself, wrong. Causing pain to another person is, other things being equal, wrong. Even taking a small sample of blood from a person's arm has to be done with a justifying reason. But, in many medical procedures, the end really does justify the means – so long as it is necessary, and proportionate, and steps have been taken to minimise harms.

What can be done? It is at least necessary to be clear about any ideologies underlying research agendas, and to be clear about the means used to achieve the ends that the ideology justifies, and to make the space from time to time to reflect on these. There can be tensions between the imperatives to advance scientific knowledge, the quest for treatment, and prevention strategies and the protection of subjects. But it is crucial to scrutinise any underlying motivating ideologies and subject them to the full light of awareness. For instance, it is fair to say that the imperative to speed in genomics is not motivated solely by an altruistic, other-regarding desire to cure disease. Much of the imperative to speed is driven by more mundane imperatives such as the competition from rival research groups and the need to rack up publications in the continual quest for the renewal of funding and contracts. This is also a pressing moral reason to be realistic about the hype surrounding genomics research, for this might act to add layers of urgency to the imperative of research.

9.9.3 How Can Responsible Behaviour Be Fostered?

Good authority figures, as well as peers, can provide a role model for responsible behaviour. In Milgram's work, seeing a peer defy the experimenter significantly increased the chances of a subject stopping the shocks early. Other work in social psychology backs this up. We are more like sheep than is comfortable to believe; but even in herds of sheep, there are a few who can sometimes show the way.

When there is a supportive community where we expect recognition and possible punishment, people are more likely to act in a responsible manner. A car was left abandoned in two very different neighbourhoods (Time 1968; Zimbardo 2007). In a quiet, middle class suburb of California, not only did the car remain undamaged, someone even closed the bonnet to protect it. In an inner city area of New York, it was cannibalised almost immediately by a succession of passing strangers. And experimental work has also shown that in large communities, without punishment for free riding, co-operative behaviour quickly deteriorates – but can be restored by the introduction of punishment for non-co-operators.

One attraction of calling for chains of responsibility in networks of researchers is the restoration of the Nuremberg call for individual responsibility that cannot be delegated away. Whilst attempting to foster this is laudable, the insights gained from phenomena such as the bystander effect suggest strongly that appointing particular people to particular responsibilities where they personally are accountable may be a necessary safeguard. At the same time, as we have seen, the unpredictability of the moral issues we may face will mean that in some cases, no one specific is charged with particular responsibility. Thus, fostering a culture of open discussions and questioning, or having specific individuals charged with a wide responsibility for listening to any ethical concerns, may be of value.

9.10 Conclusions

At the very least, institutional arrangements and safeguards must be put in place to ensure that those with scruples, or perhaps placed in particular positions that give them insight into problems that might arise ethically in the course of research, are both able and encouraged to speak out and, just as importantly, that their views have a tangible chance of leading to a change in outcome, else speaking out may actually reinforce inaction. In other words, there should be institutional support and protection for whistleblowers. The precise details of such arrangements are beyond the scope of this chapter. But the hierarchical nature of research organisations, and the different power relations within them, will set limits on the capacity of agents to act autonomously and will help shape the authoritative landscape within which they choose, think, and act. It is also crucially important that those with concerns should have some expectation and channels where possible action might realistically result from the open voicing of concerns – to avoid the problem that speaking out can act to ease consciences whilst poor practice continues.

I have argued that individuals, in acting in ways that might attract moral opprobrium, are often being pulled between different sources of moral authority. In order to come to considered judgements, close attention to competing demands needs to be made; such reflection can be hampered by the minute attention to technicalities of which Milgram spoke, minute attention that is often demanded of scientists and others. It could, however, be fostered by institutional design and practice which gave free and open space to precisely such reflection. Ideally it should include attention to how authority within institutions runs, and is distilled, dissipated, and hidden. Because of the multiple sources of moral authority, this reflection will need to encompass far more than a consideration of the requirements of formal ethics regulation. Acting well requires thought and reflection; and in order to act well, we should also reflect on the fact that we often fail to reflect well, and take steps to encourage this, both on an individual as also on an institutional level.

9.11 Study Questions

1. This example is taken from clinical genetics but has relevance to some questions in genomics research for it deals with the relevance of genetic information about one individual for related individuals. There are cases where a patient may refuse to pass on this information to relatives. A 2011 report by the UK Human Genetics Commission suggests that "The moral obligation to share personal genetic information that may be relevant in order to avoid harm may fall both on the people tested and on clinicians (or others who are aware of a relevant test result)" (Human Genetics Commission 2011). This suggests strongly that individuals may have a responsibility to share this information with family members. Likewise the General Medical Council's guidance on confidentiality from 2009 suggests that refusal to pass on genetic information to family members may justify the doctor overriding usual standards of confidentiality, again suggesting there is a responsibility to inform (General Medical Council 2009). However, social research finds that often patients do not share such information, but that routinely this is not because they are neglecting their responsibilities, but because they are taking very careful steps to manage the information, judging when, where, how, and who to tell (Arribas-Ayllon et al. 2011; Featherstone et al. 2006).

Do individuals have a responsibility to pass on their genetic information to family members?

What other responsibilities do people have that might limit or shape any such responsibility?

2. A genomics research project recruits subjects from various countries including from richer ones.

What responsibilities do members of the research project and their institutions have towards recruits from the developing countries?

Are there particular responsibilities for benefit sharing in relation to these countries that differ from responsibilities towards those from richer nations, and if so, why, what are these?

It might be argued that there are particular responsibilities to ensure that those from developing countries can get access to the fruits of any such research, such as pharmaceuticals. Do you agree?

How can this be ensured, given that the researchers may not be in a direct position to guarantee this?

Is any such responsibility owed to the research subjects only; to the particular population group only; to the whole nation from which subjects were recruited?

What about the claims of those in a neighbouring equally poor country, where subjects were not recruited?

What about the claims of individuals living in the developed nations involved in the research but who do not have adequate health insurance to access any benefits of the research in which they took part?

3. A researcher examining genomics data from a project where related individuals were recruited comes across data suggestive of possible incest. The individuals involved are all adults, and they were recruited several years ago in another country. The researcher raises the question with a colleague who thinks nothing should be done.

What are the ethical and social issues involved in such a case, and what responsibilities might the researcher have?

9.12 Coming Up Next

We have looked at many ways in which the practice of genomics research raises complex ethical questions to which we do not have ready answers. One notable feature of much current biomedical research, including genomics research, is the widespread sharing of data. This poses a number of ethical challenges including those of the extent of researcher responsibility, and the protection of privacy. It also raises the fundamental question of the balancing of ethical values, especially the welfare of persons, with the value of scientific knowledge.

References

Appelbaum PS, Roth LH, Lidz CW, Benson P, Winslade W (1987) False hopes and best data – consent to research and the therapeutic misconception. Hastings Cent Rep 17(2):20–24

Arribas-Ayllon M, Featherstone K, Atkinson P (2011) The practical ethics of genetic responsibility: non-disclosure and the autonomy of affect. Soc Theor Health 9:3–23

Austin MA (2002) Ethical issues in human genome epidemiology: a case study based on the Japanese American Family Study in Seattle, Washington. Am J Epidemiol 155(7):585–592

Berg K (2001) The ethics of benefit sharing. Clin Genet 59:240–243

Boddington P (2009) Shared responsibility agreements: causes of contention. In: Dawson A (ed) The philosophy of public health. Ashgate, Aldershot, pp 85–100

Botkin JR (2001) Protecting the privacy of family members in survey and pedigree research. JAMA 285(2):207–211. doi:10.1001/jama.285.2.207

Caparo Industries plc v Dickman (1990). 2AC 605.

Cassa CA, Schmidt B, Kohane IS, Mandl KD (2008) My sister's keeper?: genomic research and the identifiability of siblings. BMC Med Genom 1. doi:doi: 10.1186/1755-8794-1-32

Caulfield T, AL McGuire, Cho M, JA Buchanan, MM Burgess (2008a) Research ethics recommendations for whole genome research: consensus statement. PLoS Biol 6(3). doi: doi10.1371/journal.pbio.0060073

Caulfield T, McGuire AL, Cho MK, Buchanan JA, Burgess MM, Danilczyk U, Diaz CM, Fryer-Edwars K, Green SK, Hodosh MA, Juengst ET, Kaye J, Kedes L, Knoppers BM, Lemmens T, Meslin EM, Murphy J, Nussbaum RL, Otlowski M, Pullman D, Ray PN, Sugarman J, Timmons M (2008b) Research ethics recommendations for whole-genome research: consensus statement. PLoS Biol 6(3):e73, doi:10.1371/journal.pbio.0060073

Chandola T, Britton A, Brunner E, Hemingway H, Malik M, Kumari M, Badrick E, Kivimaki M, Marmot M (2008) Work stress and coronary heart disease: what are the mechanisms? Eur Heart J 29:640–648. doi:doi:10.1093/eurheart/ehm584

Cho M (2008) Understanding incidental findings in the context of genetics and genomics. J Law Med Ethics 36:280–285

Chokshi DA, Kwiatkowski DP (2005) Ethical challenges of genomic epidemiology in developing countries. Genomics Soc Policy 1(1):1–15

Chokshi DA, Parker M, Kwiatkowski DP (2006) Data sharing and intellectual property in a genomic epidemiology Network: policies for large-scale research collaboration. Bull World Health Organ 84(5):382–387

Darley JM, Latane B (1968) Bystander intervention in emergencies: diffusion of responsibility. J Pers Soc Psychol 8(4):377–383

Donaldson SL (2003) Making amends: a consultation paper setting out proposals for reforming the approach to clinical negligence in the NHS. Department of Health, London

Featherstone K, Atkinson P, Bharadwaj A, Clarke A (2006) Risky relations: family, kinship and the new genetics. Berg, Oxford

Fitzpatrick M (2001) The tyranny of health: doctors and the regulation of lifestyle. Routledge, Abingdon

Forsberg JS, Hansson MG, Eriksson S (2009) Changing perspectives in biobank research: from individual rights to concerns about public health regarding the return of results. Eur J Hum Genet 17(12):1544–9

General Medical Council (2006) Good medical practice. General Medical Council, London

General Medical Council (2009) Confidentiality. General Medical Council, London

Glover J (1970) Responsibility. Routledge and Kegan Paul, London

HapMap Consortium TI (2004) Integrating ethics and science in the International HapMap project. Nat Rev Genet 5:467–475

Human Genetics Commission (2011) Increasing options, informing choice. Human Genetics Commission, London

Johnson C, Kaye J (2004) Does the UK Biobank have a legal obligation to feedback individual findings to participants? Med Law Rev 12:239–267

Knoppers BM (2000) Population genetics and benefit sharing. Commun Genet 3:212–214

Mascalzoni D, Hicks A, Pramstaller P, Wjst M (2008) Informed consent in the genomics era. PLoS Med 5(9):e192

Mason JK, MCall Smith RA, Laurie G (2002) Law and medical ethics, 6th edn. Butterworths, London

McGuire AL, Caulfield T, Cho MK (2008) Research ethics and the challenge of whole-genome sequencing. Nat Rev Genet 9(2):152–156

Milgram S (1974) Obedience to authority. Harper Collins, New York

Miller FG, Mello MM, Joffe S (2008) Incidental findings in human subjects research: what do investigators owe research participants? J Law Med Ethics 36(2):271–279

Moriarty T (1975) Crime, commitment, and the responsive bystander: two field experiments. J Pers Soc Psychol 31(2):370–376
National Institute for Clinical Excellence. http://www.nice.org.uk/. Accessed July 2011
New York Times (1964) Queens woman is stabbed to death in front of her home. New York Times, March 13, p 26
Re Creutzfeldt-Jakob Disease litigation (2000).
Rotimi C, Leppart M, Matsuda I, Zeng C, Zhang H, Adebamowo C, Ajayi I, Aniagwu T, Dixon M, Fukushima Y, Macer D, Marshall P, Nkwodimmah C, Peiffer A, Royal C, Eiko S, Zhao H, Wang VO, Mcewan J, HapMap Consortium TI (2007) Community engagement and informed consent in the International HapMap project. Commun Genet 10:186–198. doi:10.1159/000101761
Sachs L (1996) Causality, responsibility and blame – core issues in the cultural construction and subtext of prevention. Sociol Health Illn 18(5):632–652
Sartre J-P (1973) Existentialism and humanism (trans: Mairet P). Eyre Methuen, London
Schroeder D (2007) Benefit sharing: it's time for a definition. J Med Ethics 33(4):205–209
Stansfeld SA, Marmot M (eds) (2002) Stress and the heart: psychosocial pathways to coronary heart disease. BMJ Books, London
Time (1968) Diary of a vandalised car. Time Magazine, October 1st, Trials of War Criminals before the Nuremberg Military Tribunals under Control Council Law No. 10, vol. 2 (1949). US Government Printing Office, Washington DC
Warin M, Moore V, Zivkovic T, Davies M (2011) Telescoping the origins of obesity to women's bodies: How gender inequalities are being squeezed out of Barker's hypothesis. Ann Hum Biol 38(4):453–460. doi:doi:10.3109/03014460.2011.591829
Warnock M (1990) Boys and girls to order, and rightly. The Times, April 20th
Zimbardo P (2007) The Lucifer effect: how good people turn evil. Rider, London

Chapter 10
Data Sharing in Genomics

'Open discussion of ideas and full disclosure of supporting facts are the bedrock for scientific discourse and new developments,'

(Toronto International Data Release Authors 2009)

Abstract This chapter looks at how data sharing practices in genetics give rise to ethical issues for both the researcher and the participant. Like some of the previous discussion in the book, it demonstrates how an understanding of the social practice of science is needed to appreciate and analyse fully the ethical issues that arise. Questions considered include the acknowledgement of contributions to scientific work and publications, which requires a discussion of the career structure of researchers. Discussion of ethical issues arising from data sharing also involves analysis of notions of ownership of information, public and private work in science, intellectual property, and Enlightenment values of science and scientific knowledge. Also discussed are questions of privacy and confidentiality of data; consent to data sharing; and the use of archived data.

10.1 Introduction

Data sharing in science has a long history. In genetics in particular, the sharing of data and samples in the spirit of scientific collaboration and to expedite the speed of progress has many roots and has been occurring since the early stages of this branch of biology. In genomics, sharing data received a significant push from the race to sequence the human genome. In the face of fears that the private consortium, Celera, would seek to patent portions of the human DNA sequence, the international public effort developed the technique of publishing sequence data on a daily basis as soon as it was available to obviate the chance of patenting (Davies 2001). This ethos has added weight to widespread support for the sharing of genomics data, such that sharing data is now seen as the norm by major funders of such

P. Boddington, *Ethical Challenges in Genomics Research*,

DOI 10.1007/978-3-642-23699-0_10,

research; it also fed into the almost ubiquitous trope of speed in genomics research. A great deal of genomics data is available for free use by researchers worldwide.

There are many scientific reasons to share data: both to achieve large samples to produce more robust results, and to compare different population groups. Including diverse populations may be done for reasons of equity as well as for scientific completeness. Furthermore, sharing data may occur within scientific collaborations where expertise and technological capacity are shared and efficient use can be made of existing resources and minimise demands on data subjects. Hence it is easy to make the simple claim that to share is to be good; to withhold data puts one in the camp of those who want, in this frontier land of research, to claim for themselves a portion of what seems emblematically to belong to the whole of humanity: our DNA. But things are not that simple; and even if practitioners in the field of genomics remain on the whole committed to the value of sharing data, the foundations of this value stand in need of further articulation.

There is continuing development of data sharing policy and practice in genomics, as well as in other areas of biomedical research and biobanking. International meetings have developed policies with the involvement of key international players. The need for global agreement is of course underscored by the international nature of much research collaboration in genomics and elsewhere in science. These international meetings acknowledged the complexity of the issues. As a result, in current policy in genomics research, not all data is simply released in a free-for-all, because of the different interests of those involved, although there are also projects underway with a strong commitment to open access which publish all results freely, for example the Personal Genomes Project.

A crucial issue is the nature of the data itself as well as the conditions of its production. Genomics data is produced by particular researchers, funded by particular funders, and importantly, derived from particular individuals and in some cases, particular disease or population groups. We shall also see how the ways in which genomics information is conceptualised or imagined play a part in attitudes towards sharing genomics data.

Part of what these policy documents have done is to recognise the different yet valid interests of a number of different parties. There are different values at issue in data sharing, which are not necessarily well-articulated and which may conflict. These include: values associated with the possession of knowledge; with the obtaining of such knowledge; with the transmission of such knowledge; with the advancement of science; as well as particular issues concerning knowledge derived from human subjects. To some extent, it is possible to separate these out into scientific, policy, economic, and ethical reasons. However, what one might initially take to be a sound scientific reason to share data may turn out on reflection to be driven by certain values or by contingent facts about the way in which science is currently conducted. Many reasons given relate to notions of sound or good scientific practice, and ideas about what constitutes such good practice invariably include values which relate not just to the production of reliable scientific knowledge but to the operation of healthy, thriving scientific communities, and of the characteristics of individuals within these communities.

One way of understanding some of the problems that arise is by considering two perspectives on genomics knowledge. One what might be called a bird's eye perspective, genomics is not perhaps as has been thought, "the" key to understanding the core of our species, but certainly a major component of this understanding. This is knowledge that encompasses a great deal of individual and group variation, but which nonetheless has great commonalities across our species and vast implications for us all. From this approach, genomics knowledge looks like an iconic candidate for the status of knowledge that belongs to all of us – knowledge that is often referred to, significantly trading on the notion of genetic inheritance, as "our shared human heritage".

But there is another perspective from which genomics knowledge is understood – as deeply personal knowledge that goes to the heart of individual identity, and over which the individual from whom it derives has a strong claim of control. These two perspectives may collide where sharing genomics data produces challenges for the protection of the individuals from whom the data derives. Policy on sharing data universally recognises this problem – even for projects practicing full open access which do this in conscious recognition of the differences of opinion on this issue – although the precise nature of the solution, and where the balance is struck, varies. But at a more theoretical level, this tension between the shared, public nature of scientific knowledge and the view of (some) knowledge as private and personal opens up fascinating questions of why we value knowledge, why we value sharing knowledge, and the openness of scientific inquiry.

As the "central dogma" of genetics holds, biologically there is a flow of information such that DNA through RNA codes for proteins, the building blocks of biological organisms, so too can we understand the debates about human access to and control of genetic information as coding for many of the central issues of ethics – the relation between the self and the material world; and the relation between the individual and others, both others known and the generalised unknown other, including those long dead and those yet to be born. Crucially too, these debates raise the important issue of the relation between ethical values and other normative values, namely those associated with the value and production of knowledge; and in doing so raise questions about the nature of human welfare, human flourishing, and what it is to live well.

Sharing Knowledge, Sharing Data

There are strong links between questions of the sharing of knowledge and the sharing of data but it is important to note that the issues involved may also be different.

We can think of data as the raw ingredients from which knowledge is constructed.

Arguments in favour of sharing knowledge may depend upon the type of knowledge it is and what it is knowledge about. We may distinguish, for instance, between public knowledge, private knowledge, and generalisable scientific findings.

(continued)

Arguments about sharing data may differ from arguments about sharing knowledge especially when the same general knowledge may be derived from different data sources. We all stand in the same relation to generalisable scientific knowledge: but data is specific to the particular individuals and groups from whom it is derived. In debating the exchange of data, the nature of the data source(s) comes more to the fore, although these issues will be of relevance to sharing knowledge especially where particular knowledge has been specifically derived from certain data sources.

There may be arguments to share data based upon, for example, the difficulty of accessing the data, and other general efficiency considerations. In the case of data relating to individuals and groups, there may be particular considerations for or against sharing that refer to the production of that data from these individuals or groups or the content of the data. These considerations may also apply to generalised knowledge about these individuals or groups; but given that this knowledge may be derived from different data sets, these considerations may not exactly coincide.

Arguments that knowledge about the human genome "belongs" to mankind in a strong sense may in general be more robust than arguments that any particular pieces of data belongs to humankind.

10.2 Reasons to Share Genomics Data: Views from Policy and Practice

Our discussion of the values underlying data sharing policies and practice begins by discussing various policy documents dealing with data sharing in science in general and in genomics in particular. These include documents from the Wellcome Trust, the National Institutes of Health, the Medical Research Council, the Organisation for Economic Co-operation and Development, the Biotechnology and Biomedical Sciences Research Council, and the cancer Bioinformatics Grid (caBIG®). The following section provides a succinct summary and discussion of some of the recurring themes found in data sharing policy, supplemented by discussion with those involved in data sharing in various guises in practice.

10.2.1 Data Sharing as the Norm

There is a broad consensus that data sharing is not only of great value but should be seen as the norm. It is now the case that many funders of scientific research require data sharing as a condition of funding for many of their projects, or require that where data is not shared, an explanation for this must be provided.

10.2.2 Improving Scientific Outputs

Data sharing is commonly held to maximise research outputs, and to lead to improvements in the quality of research, including improvements in research practice. Data sharing is said to reinforce open scientific inquiry; to increase the value of international cooperation; to allow for the validation of results; and to improve and open up pathways for dissemination of findings.

10.2.3 Fostering Novelty and Scientific Creativity

There are claims that data sharing can lead to innovation, to novel benefits, to new opportunities for advancement to the goal of health improvement, to novel areas of work and topics not anticipated by the original investigators, to the testing of novel hypotheses and analysis methods, and can allow the creation of new data sets as data are linked and pooled.

10.2.4 The Value of Sharing Certain Sorts of Data

Specific reasons to share certain types of data are sometimes given: there is value in sharing data which is unique; not possessed by anyone else; or hard to replicate. It is also the case that immense volumes of data are generated in some areas which outstrip the capacity of any one individual or even any one team to perform all possible analyses, again giving reasons to share to enable fuller analyses.

10.2.5 Visions of Scientific and of Human Progress

A notion of a vast community stretching over time and around the world to construct the edifice of scientific knowledge is embedded in the heart of much of the ideology of data sharing. The OECD explicitly claims that the exchange of ideas, knowledge, and data is fundamental to *human* progress and as such a core part of OECD values. Talk of reasons to share is oriented to the future progress, and is often tied to plans for assuring the long term sustainability of sharing practices.

But note too that the OECD also sees one value of data sharing as encouraging diversity of opinion and of studies (p. 3). We will see later that the desire for building a harmonious community of data sharers who share common values and which discriminates amongst those who are and are not allowed into the community aims at a certain moral conformity which fits uneasily with this perceived need for a scientific diversity.

10.2.6 The Value of Data and the Value of Knowledge

Some data sharing policy statements explicitly address the relative value of data and of knowledge generated from that data. The Wellcome Trust policy statement on data sharing explicitly regards research data as "a fundamental research output"; such a claim is closely linked with drives to assure adequate and appropriate recognition for the work of those who generate data (see below on scientific practice).

10.2.7 Data Sharing and Good Scientific Practice

The benefits of sharing data are also intimately linked to notions of what constitutes or is seen to be good scientific practice. Moreover, in many policy documents, encouraging the sharing of data is seen as good scientific practice; good scientific practice is considered an essential to allow data sharing for general benefit; but it is further argued that one benefit of sharing data is this very fostering of improvements in good scientific practice.

10.2.8 Good Data Sharing and Stewardship

There is broad agreement that good scientific practice involves taking steps to ensure good stewardship of data for long term use, and for harmonisation of standards to allow international cooperation and sharing; and also that data sharing and its requirements must be considered at the outset of a project.

10.2.9 Data Sharing and Research Participants

Many point out that sharing data can reduce the burdens on participants, by making most efficient use of data already generated.

10.2.10 Data Sharing Brings Benefit to the Public

There are frequent references to the benefits for the public in sharing data. But what might be meant by "benefits to the public"? One interpretation is an abstract ideal that objective scientific knowledge belongs in the public domain. Another more pragmatic interpretation is that data sharing contributes to public good through the production of scientific knowledge which can be translated into health and other benefits accessible to all. Yet another quite different reading of public benefit however is of getting one's money's worth; the public funds much research through taxes. "The public" may be to that extent confined to a particular group, for example a nation of tax-payers, rather than humanity as a whole.

10.2.11 The Timing of Data Sharing, Speed and Brakes

Discussions about the timing of data sharing are fascinating. There is broad agreement that data should be released in a "timely" manner: that "data delayed is data denied", implying that delays in data sharing are equivalent to withholding data, and that the value of data is often linked to its timeliness. Science is an activity, and its intimate link with speed is thus as so often demonstrated. Any delays must be justified, and must be minimal.

10.2.12 Interests in Controlling Data

There is a vital proviso to timely data release, because there is also universal agreement that those who produce the data have a "reasonable" and "legitimate" interest in getting first go at data analysis and the production of publications derived from the data, or to obtain patents or otherwise protect intellectual property (IP). It is an embedded part of scientific practice that as well as the rewards of recognition by one's peers, the rewards of publication are necessary to career advancement, and indeed often for simply maintaining a career at all. This is however surely a contingent fact about the current practice of science, since it is quite possible to imagine that good researchers could be awarded continuous employment without the current constant demands of producing publications. The argument that researchers' interests can justify a break in the progress of research serves to remind us of the multiple sources of the imperative to speed with which talk of genomics is suffused.

International Data Release and Data Sharing Agreements

The Bermuda Principles (International Strategy Meeting on Human Genome Sequencing 1997, **1996)**

The First International Strategy Meeting on Human Genome Sequencing met in Bermuda in 1996. Participants included the Wellcome Trust, the UK Medical Research Council, the National Institute of Health (NIH)'s National Centre for Human Genome Research, the Human Genome Organisation (HUGO), the German Human Genome Programme, and the Human Genome Project of Japan. Principles agreed: the free release of primary genomic sequence data into the public domain, in order to encourage research and development and maximise social benefits; the rapid release of all human genome sequence data generated by large scale sequencing centres with public funding, to prevent any such centres establishing a privileged position. At the 1997 meeting sequence quality standards were agreed. Standards for addressing discrepancies in data, for archiving data and for standardising sequence annotation were proposed. A discussion of the etiquette of claims over the mapping of particular regions of the genome suggested that potential conflicts be resolved by early agreement.

The Fort Lauderdale Data Sharing Agreement: (The Wellcome Trust 2003)

The Wellcome Trust organised this meeting at which a tripartite sharing of responsibility between funding agencies, resource producers, and resources users was agreed.

This meeting endorsed the Bermuda Principles, and extended the principle of rapid pre-publication release to data from large scale "community resource projects" specifically devised to create sets of data, such as the SNP consortium and the International HapMap Consortium (HapMap Consortium).

The meeting recognised that pre-publication data release might conflict with the incentive of publishing the first analysis of one's own data, and stated that the contributions and interests of data producers should be respected and supported by users of the data and by funders.

Funding agencies should require free and unrestricted data release from community resource projects as a condition of funding, and encourage involvement in such projects to serve the community. Resource users were receiving great benefits from larger, richer, and high quality data and should provide correct citations for data used and respect the legitimate interests of the data producers, for example in prominent publications. The development of incentives to encourage the sharing of data from other projects was to be encouraged. Occasional violation of scientific etiquette in the use of these resources was seen as a risk to be borne for the larger benefits to be gained. The value of maximising public benefit was used as a ground for these principles.

The Toronto International Data Release Agreement (Toronto International Data Release Authors 2009)

This multinational and multidisciplinary group aimed to reaffirm the principles of early data release for a wide range of data sets with a wider group of stakeholders, and to develop a set of suggested "best practices" for funders, data producers, data users, and for scientists in their roles as editors and reviewers. Issues covered including the simultaneous release of meta-data that would enable users to maximise use of the data, and concerns of confidentiality and privacy with clinical data. Rapid pre-publication release of data from large reference data generating projects in which the community has an interest is to be encouraged, forming part of the original set up of such projects, although not mandated for more hypothesis-driven projects.

Values and concerns mentioned include fair credit to investigators for data generation and release, community interest in resource use, the conditions of co-operative collaboration including transparency and open dialogue, as well as attention to data quality and standard setting, and the encouragement of debate. Possible problems with publications were recognised but seen as weighing against "the much greater overall benefit of early data release".

10.3 Data Sharing and Scientific Practice

The intimate link of practice with scientific output is present in the claim of the OECD: "The progress of science, however, depends on more than just technologies. Research practices, policies, support systems and cultural values all affect the nature of new discoveries, the rate at which they are made, and the degree to which they are accessible and used" (p. 9). But what is "good scientific practice"? (Resnik 2007) Changing practices of data sharing itself shows that what should or should not constitute scientific practice are changing rapidly (Kaye et al. 2009).

Good scientific practice includes the notion of a scientific community, and also explicit recognition that different members are differently placed within this community, and may be differently placed in relation to the practice of sharing data. The implications of these different placements may yet need fuller consideration. An examination of these issues is crucially important because practices of data sharing may themselves help to shape and change the nature of scientific communities and different actors may stand to gain or lose substantially as a result of the sharing or withholding of data. It is also important to realise that there may be inadvertent repercussions of data sharing policies and practices.

There are also potential conflicts. In some accounts of good practice in data sharing, there is an ideal of reciprocity: good data sharers are those who can make best use of the data, or who also have data to share. In other words, the higher your baseline of resources, the better data sharer you would be able to be; there is a danger then that data sharing policies and practices might establish an elite "club" of well resourced, well behaved data sharers. But on the other hand there are many who are concerned that issues of global inequality be addressed where they can through data sharing, for example in agreements that build capacity in developing countries in both infrastructure, staff, and training.

Another source of conflict and difficulty concerns calls for harmonisation of standards in producing and presenting data. Whilst this is an understandably desirable goal if data are to be shared globally, there is a danger that harmonisation may become homogenisation. Not only might this lead to a drop in diversity within scientific practice, it may also lead to problems for those who do not comply with such harmonisation standards for whatever reason, who may then find themselves excluded from data sharing practices.

10.3.1 Stewardship of Data

Good scientific practice may include notions of good stewardship of data and samples, but "stewardship" is a term with invariably positive connotations which means little unless its application is fleshed out. Good stewardship of data may be thought to imply adhering very strictly to the letter of informed consent agreements with research subjects – generally resulting in more restrictive sharing of data; but there are those who consider that the interests of research subjects are best served by maximising use of their data and samples. Some researchers feel that stewardship responsibilities involve drawing upon their long term relationship and understanding with a group of subjects (for example in longitudinal or cohort studies), which gives them a certain license or authority to interpret their wishes and to understand what it is to act in the best interests of the research subjects (Vickers 2006; Delamothe 1996).

Stewardship of data may involve allowing access to it and may involve withholding data. It may involve judgements about who is a good custodian of the data and who will make good use of it. It is instructive briefly to consider relevant findings from research into the communication of genetic (and other medical) information within the family. Here those with stewardship over the genetic information may give great thought to issues such as when to tell, how much to tell, who to tell, and for what reasons (Arribas-Ayllon et al. 2011; Featherstone et al. 2006). These may be precise, anguished, and nuanced judgements that require sensitivity and insider or local knowledge. These judgements may conflict with external judgements, for example from members of the medical profession about what appropriate moral rules and rights there are with respect to the sharing of genetic information.

10.3.2 Rewards, Sanctions, and the Scientific Community

Good scientific practice will involve appropriate rewards for data sharing and stewardship and sanctions for inappropriate practice. But much hangs on what these rewards and sanctions are. For example, there are current suggestions of developing a notion of who counts as a "*bona fide*" researcher for the purposes of validating data users, and this is also accompanied by suggestions that those failing to meet standards might be excluded from access to data resources. There are also moves to ensure that those who produce data are adequately rewarded and recognised, through appropriate acknowledgements on publications, and in terms of how universities and funders reward achievements and renew contracts. However, it is also important to realise how embedded these are in the particularities of social, economic, and cultural circumstances. The OECD points out that rewards for sharing data are linked to cultural values (p. 12).

Given that, as we have seen, there are different notions of what might constitute good use of data, good stewardship of data, or valid reasons not to share data, together with warnings that the research community may contain inequalities that data sharing practices may reduce, or yet may increase, there are real dangers attached to a system of sanctions that may leave some researchers or research communities outside of the global system of accessible research data, and possibly barred from publication in certain journals.

The importance of this issue can scarcely be overstated; ideals of free data sharing, inspired at least partially by the notion that knowledge of the human genome should be freely available to humanity, are potentially in danger of locking some outside the community of scholars and researchers. If this is to be done, it should at least be done with great care. This is one reason why careful attention must be paid to what is expected of researchers, and to how the "good researcher" is characterised (Redner 1987). These issues also of course relate intimately to the question of the responsibility of researchers discussed in the last chapter.

Open Access Publishing

Moves to "open access publishing" can be seen as part of a shift to more democratic and equal access to knowledge. Traditionally, the journals that publish scientific and other academic papers have only been available under subscription and often available only in institutional libraries closed to those outside of the institution. Such journals may be available online but available only on, for example, a "pay per view" basis, taking this beyond the means of many. Some journals have part of their content freely available online, for example, allowing access to certain key papers or to content prior to a certain date or providing abstracts and opening portions of papers only.

Open access publishing can be hailed as the democratisation of knowledge. But it should be noted that there is usually a fee for publishing in such journals. Costs are met not by the readers but by the contributors (or usually, by their institutions). Hence, it is not open to all to publish in such journals as many cannot afford the substantial fees. Even those working together in the same institutions may have different funding constraints that may bar them from access to such "open access" journals. Neither are such journals open for those with shared, intermittent, or otherwise unreliable internet access.

(continued)

(This is one of the many considerations behind moves to share benefits and improve research infrastructure in developing countries.) Other rather different considerations of openness in publishing relate to the process of peer review and debates about its fairness.

Why not just post results online if one wants free access? Some have suggested that this will allow free assessment of work and greater exchange of information. On the other hand, such assessment is highly complex, specialised, and time consuming work; if editors and reviewers do not do this for the ordinary reader, the weight of available work of unreliable quality may make sifting through it a nightmare. Furthermore, the stamp of endorsement of publishing in high quality journals is at present a key criterion for employment and promotion.

10.4 The Value of Knowledge and the Value of Sharing Knowledge

Having looked at reasons given for and against data sharing in policy arguments we now turn to examine background philosophical arguments about the value of sharing knowledge and data. We looked in Sect. 3.2.4 at the relation between ethical values and the value of knowledge and of scientific progress, and at the difficulties of characterising why we value knowledge as we do. Why do we value sharing knowledge? Why should we share data, the raw building blocks of knowledge? Data sharing is valued insofar as it helps to advance the progress of knowledge. But why do we value knowledge? In medical science, and in genomics, knowledge can be valued for its ultimate application in the service of improvements to human health and to life more generally. There are economic benefits of knowledge, ranging from the profits from intellectual property to those producing pharmaceuticals and medical devices, to the knock-on effects on the economy of improved health as well as the internal economy of the health care industry. But knowledge is valued not just instrumentally, but also for its own sake. Genomic knowledge is valued not just for its potential medical application but also for its interest in showing who we are and how we relate to others, as the trade in genetic ancestry information reveals; there is undoubtedly for many people a narcissistic element of self-love in discovering minute details of one's own genome, as well as an element of simply scientific fun: you can find out, after all, if you can smell asparagus in your own urine, and most people already know whether or not they like broccoli; for many, it's a form of amusement to have these trivial facts confirmed in a direct-to-consumer DNA analysis.

But if you ask more directly why we value knowledge over and above its usefulness, it turns out to be hard to produce an articulated answer. We seem to be creatures who just like finding out stuff, as the quotation from Aristotle in Sect 3.2.4 illustrates, creatures with an inherent curiosity. As well as institutions for learning, continuing education classes, billions of Google searches, and pub quiz nights throughout the land, all testify to our insatiable appetite for knowledge *for its own sake*.

We can find a number of different reasons for sharing knowledge. There is sometimes a childishly simple motivation: for the thrill of being accredited with its discovery. We share knowledge often because of our nature as show-offs, the urge to be the first with the news. Leaving aside the noble pursuit of benefits to the sick, fear of being beaten to publication is a prime motivator in science and a prime reason behind the constant rhetoric of speed.

Consideration of scientific credit leads us to be aware of further questions. What follows from the attribution of credit for knowledge? Should there be particular rewards and acknowledgements tied to particular knowledge achievements? And, in the context of data sharing, and policies that may aim to emphasise the value of producing data, we can ask the important question of whether or not the same credit should accrue to producing and possessing data as they might to producing and possessing knowledge. Credit may accrue to producing knowledge, possessing knowledge, curating knowledge, and transmitting knowledge.

Further to this, the impetus to share data derives at least in part from the nature of the scientific enterprise, indeed from the nature of science: there is an awful lot of it. It is just outside the capabilities of individuals acting alone to further science very much; each of us, whether as layperson or scientist, builds our picture of the world upon testimony, a term used in epistemology to mean knowledge known second hand through the knowledge of others. In science these others stretch around the globe (and beyond the globe, if you include the knowledge gathered by astronauts) and far back in time. If we are to advance science, we have to do it together, and this means that sharing data will help enormously with our endeavours.

There is more to sharing knowledge than simply advancing the goals of a scientific community. It is also a strong grounded notion that scientific knowledge is the province of everyone: that it is public, it belongs to us all, in the same way as other public goods such as the air we breathe and the moon and sky at night: it is part of our common human heritage. This assumption is present behind much of the thinking supportive of data sharing in science.

What vision of the world, and of our scientific understanding of it, lies behind the notion that scientific knowledge belongs to us all, and should be public knowledge? Implicit in such a notion perhaps is an idea related to the objectivity of scientific knowledge. One way of expressing this is that it is universal or eternal knowledge that is the same seen from any point of view, and hence belongs not to one person more than to any other. An argument can then be made from the claimed objectivity of knowledge to a democratising of it: objective scientific knowledge stands in no special relation to any one individual. (This can then be contrasted with other forms of knowledge which are private or limited to certain individuals; watch this space for forthcoming trouble.)

Additional reasons to share certain types of knowledge derive from their subject matter. It was Alexander Pope who said "The proper study of mankind is man"

(Pope 1796): it seems particularly plausible to claim that scientific study of humanity should share its findings dispassionately with its subjects: all of us. It is "about" us (See Sects. 2.2.4 and 5.2.7).

A degree of repugnance at the idea of private control over the genome is felt even by many who are in general terms otherwise content with notions of private and intellectual property. Abhorrence at the notion that someone could have private control over "our" genome goes very deep, and forms an enormously strong impetus behind the continuing drive to share data as well as resistance to patenting aspects of the human genome.

Ideas of the public nature of scientific knowledge of the world came to the fore with the Enlightenment, sweeping aside notions of scientific secrecy, and linking this openness of knowledge with an opening up of society, which included the passage towards democracy and the increasing value attached to the individual within the open society. Public access to scientific knowledge is therefore at once a democratisation of knowledge. Democracy, the value of the individual, and the free pursuit of science are mutually reinforcing, insofar as knowledge can empower individuals and free them from unnecessary dependency on those with control over sources of knowledge.

But we can ask what is it for knowledge to be public knowledge. Given the immense complexity of much science, high levels of educational achievement are often needed "fully" to grasp it (although whether our human brains will ever "fully" grasp the whole nature of scientific reality is a moot point. A physics professor told me over tea once that he hoped string theory didn't turn out to be true "because the maths are so darned complicated"). But even given reasonable levels of education, allowing access to reasonable levels of comprehension of science, much has still not been accessible to many. This of course relates to questions of education and to the issue of informed consent to research in complex areas of inquiry including genomics.

If knowledge is derived from data originating in an individual or group, then an argument can be made that the individual or group has a particular claim to access to knowledge produced from that data (or sometimes, simply to access to that data). There are disagreements on this point; partly these relate to the difficult logistical issues and financial implications of feeding back such information to research recruits. But the general notion that a subject has a particular claim upon any knowledge or data that derives from that subject is a common one that merits further examination and articulation. It is not necessarily a claim that will stand up to close scrutiny.

A further basis for justifying claims that data or knowledge should be shared is the simple "we paid for it" argument. The ideas that the public paid for government sponsored work, and therefore should both have access to its results, and to the fruits of the application of this work, is a strong one. Note well, however, some important points. The public who pays for research and its considerable infrastructure is not necessarily equivalent to the public who potentially stands to benefit; and the public who pays for this research certainly is not the global mass of humanity as a whole. Furthermore, the "we paid for it" argument is held equally strongly by the commercial sector. There are particular issues with the many public-private joint research projects; many of these are limited in what data they are able to share because of commercial clauses restricting data sharing, or restricting information about the directions of research. It's important to remember such difficulties especially in the light of calls to blacklist those who do not or will not share data.

Valuing Knowledge, Valuing Data

Philosophers since Socrates have grappled with the question of why we value knowledge, and what knowledge actually is (Plato 1956). It's recognised that there is a distinction between knowledge and "mere" true belief: a true belief might be a lucky guess, for instance, or a fluke (as when one misreads false information to get an accidentally true answer). There is still no philosophical consensus about how knowledge should be defined, but there is broad consensus that knowledge should be gained in the right way; true knowledge can claim to be *justified*. The question of what this "right way" to gain knowledge might be exactly still keeps philosophers busy.

One approach to answering this question why we value knowledge over mere belief involves looking at the method through which knowledge is generated; if the method is robust, then true knowledge will result (rather than mere belief). Likewise, we have seen that a great part of the policies on data sharing in science have been concerned with what constitutes good scientific practice. Sharing data is seen to be part of such practice for many reasons but intimately because of its role in producing and validating scientific knowledge. The practices enhanced by sharing data, such as validation of results, enlarged pools of data sets, and scientific collaboration and communication, are part and parcel of the practices of justification that constitute which of our beliefs count as knowledge.

Hence, in answer to the question of whether or not we value data only because it leads to knowledge, we can reply that in a significant respect, to value one is to value the other.

10.5 Reasons Not to Share: Privacy and the Control of Personal Information

We need also to ask what vision, what understanding, lies behind the notion that information should not be shared. This may be for a number of reasons, many of which are implicit as antitheses of some of the dimensions of the value of sharing knowledge articulated above.

Biomedical knowledge presents particular issues in the opposition between generalisable knowledge relating to humanity as a whole, and personal private knowledge that is the concern and province of the individual (and sometimes of his or her doctor) alone. Genomics knowledge is arguably a particularly acute exemplar of this tension, even if this tension owes much of its force to imagined properties of genetic and genomic knowledge that give it greater salience than it may on fuller reflection merit. But these imagined notions can have a powerful sway over our moral judgements even if they are not fully justified.

A different but linked claim is that personal genetic information "belongs" to the individual concerned; it's "their" information. Such claims are often closely linked to notions of autonomy, and sometimes to notions of respect for persons. In order to respect a person's autonomous control over their self, it is often argued, we must allow them control over the uses and access to genetic (or other) information relating to them. There are of course great complexities with any arguments drawing upon the notion of autonomy especially in relation to genetic information as we saw in Chap. 6.

There is an interesting and important sense in which claims that individuals have particular rights to control their personal genetic data echo claims that researchers who produce data should be accorded particular rights of access to it, for example, the time-limited rights to publish material based upon that data. The data is derived from the data subject in a material physical sense; and also from the "work and skill" of the researchers (see Sect. 10.6). Where data about an individual can only be created via the skilled efforts of another, as is the case with genomics data, there is automatically more than one party to the potentially competing claims about rights of control and access. There is obviously therefore scope for conflict here. So we can see here yet another practical ethical question relates to abstract conceptual arguments, in this case about rights.

An obvious argument claims that individuals should have particular rights of control over personal genetic information because of the need to protect their privacy, and because of possible harms that might befall them should the information be misused, for example in discrimination concerning employment or insurance, or in social stigmatisation (Church et al. 2009). Such claims can be based upon the argument that individuals are best placed to make such judgments of the need to manage such possible harms; and/or on the need to respect individual rights created by consent agreements and other laws and regulations governing privacy and confidentiality. Against this, researchers could claim that adequate protection to privacy and possible harm can be given, and that they have particular grounded claims to the free use (including sharing) of data to advance science and indeed their own careers.

There is a substantive practical point that arises from different conceptions of what might ground individual claims over control of personal information. For those who argue that such control is justified as an effective strategy of minimising the harm that might result to individuals from infringement of privacy, an appropriate response might be to seek to protect individual information by data security measures including anonymisation and the use of aggregated data. However, if one considers that rights of individual autonomy, or respect for persons, implies that individuals have right of control over the use of personal data, then such techniques are not enough: even if an individual's data is pooled with many thousands of others, and even if it is vanishingly unlikely that the individual would suffer harm as a result, it is still being used; and there are those who would argue that this violates the proper control that an autonomous individual should have over their personal data.

A separate range of questions relates to questions about individual or group rights of control over genetic material of which they are the source.

10.6 Do We Own Our Genetic Information?

Ethical discussion about rights of control over genetic (or other medical) information often, either implicitly or explicitly, trades on the assumption that individuals own their genetic or genomic information. In fact such claims are very hard to substantiate; there is no basis in current property law for the notion that individuals have any legal ownership over their genetic information; the rights and interests of individuals over such information are protected by laws concerning privacy and confidentiality rather than ownership (Witte and Have 1997; Moore 2000; Spinello 2004). Moral claims for ownership of medical or genetic information also skate on thin ice.

Property rights in fact constitute a bundle of different rights usually held over resources that are finite and that therefore cannot be shared without limit. It is often assumed that to have property rights over something is to have absolute rights of control and of disposal, but quick reflection reminds us of the many limitations on property ownership, ranging from the need for planning permission to build on land one owns, possible legal challenges to wills, restrictions on activities on one's land which impact on certain species of plant or animal, and so on. Property may be private or collective, held in common for a group of people or for the whole community. There are many divergent accounts too of the philosophical basis of property rights, ranging across the whole political spectrum, from those placing great emphasis on the importance of private property, to those such as anarchists who argue against the whole institution of private property.

One highly influential account of the philosophical basis of property ownership is that of John Locke, the eighteenth century British philosopher, whose views have been significant not just historically but in shaping contemporary legal and political views (Locke 1689). Locke envisaged how property rights might have come into being in an original state of early human development before the institution of property existed. Locke's view starts out from the premise that each has control and "ownership" over his or her own person. An individual could stake a legitimate claim to a portion of land by having worked the land; by "mixing his labour" with the land. Even so, this does not give individuals rights to lay claim to more than they need, given that there is a limited amount to go around.

One might take such a view to imply that individuals therefore own their bodies, and hence that they maintain ownership rights in the cells and DNA samples derived from these. However, although these are derived from an individual's body, they are of no use in scientific research unless others have "mixed their labour" with them; this requires "work and skill" which on a Lockean account of property would give ownership rights, if to anyone, to the scientists or their employing institutions.

There have been disputes about ownership rights over the body in the cases of Henrietta Lacks (Skloot 2010) and John Moore (Moore v Regents of the University of California 1990). Immortal cell lines derived from cells from these two individuals are now in common use. Legal cases have claimed rights over the profits derived. On one view, these individuals (or their heirs) have been exploited by being deprived of these profits. On another view, Lacks and Moore did nothing to merit any share in the profits; they just happened to have suitable cells; any property rights belong to those scientists whose highly specialised efforts, together with the backing of their institutions, enabled the cell lines to be created. The case of Lacks in particular has attracted much comment, but it is well to remember that there are very many issues involved in her whole story including social and economic deprivation, levels of consent, communication, and the backdrop of issues of race; here I focus simply in the question of whether the issues involved can be illuminated by assigning property rights to cells, and questions of ownership of information derived from these cells.

It is sometimes argued from cases such as Moore and Lacks that individuals taking part in genomics research should be able to profit from the use of their information. But this overlooks the considerable differences between information, which may be shared and divided indefinitely, and physical objects; and it lacks foundation in plausible philosophical accounts of ownership. Ownership of genetic (or medical) information is even more problematic than ownership of parts of one's body, since information is not a discrete "thing" that can be limited or controlled even in the same way as a DNA sample. Once information is known, and even more so, once it is out in the public domain, it can't be returned to its owner in the same way as can an object. Nor is it considered that researchers who discover empirical facts or produce empirical data "own" these, although they may have copyright over publications. Most statements of data sharing policy explicitly consider that scientific data is something that should be released for the common benefit of researchers and of the public, yet retain a notion of a period of exclusive rights to derive publications from data for a limited period. This can be understood as an application of a Lockean notion of the privileges derived from the application of the work and skill of the data producers.

Practically too, claims that ownership rights over one's samples and genomic information ground claims to share in the profits of research overlooks the fact that tens if not hundreds of thousands of individuals are involved in genomics research; the share of any eventual profit will represent lean pickings at best. It also departs from plausible views that participation in research, especially where it involves minimal effort and risk, should be directed to the public good and towards improvements in healthcare in which we all stand to share.

10.7 The Value of Knowledge and the Value of Sharing Knowledge: Summary and Analysis

Knowledge may be valued inherently – for its own sake – or instrumentally, for the goods it can bring, and including the further progress in knowledge creation that it may bring.

The inherent value of knowledge may attach to its claims to provide a true representation of an objective world which it is in our nature to find fascinating. This in turn forms a part of a conception of what a valuable human life consists in. The notion of humanity as engaged in scientific progress is intimately linked to a major reason for sharing knowledge and data.

We may be particularly interested in knowledge that relates to ourselves, as individuals or as members of humanity, and this may be for inherent as well as instrumental reasons.

It may be argued that we have particular claims for access to knowledge or data that relates to us, either to a group of which we are a member, or to us as individuals.

The value of sharing or withholding knowledge or data has various different dimensions:

Discovery value, or the worth attached to being the first to uncover truths may underlie some of the impetus to share knowledge.

A corollary of this is that the **discoverer-producer** of this knowledge then may expect or deserve some recognition for the discovery.

A **democratic** or **universal view of truth** such that certain types of knowledge, such as scientific knowledge, are the province of everybody; we all then have a claim to access to such knowledge.

Objectivity reasons for sharing knowledge derive from the argument that we all stand in the same relation to "objective" knowledge, and that therefore it "belongs" to all of us.

Progress reasons to share knowledge relate to the vast edifice of knowledge of the natural world, which is understood as built up of parts related to the whole – sometimes as an edifice build upon foundations, and progressing to the future by building higher and higher upon these foundations; sometimes as an edifice held together in a coherent web. Sharing knowledge is therefore pragmatically necessary for the community of knowers over time to achieve scientific progress.

Humanity-content reasons for sharing knowledge relate to its subject matter: particularly knowledge that concerns humans in general.

Data-subject reasons for sharing knowledge are claims that where knowledge concerns particular individuals or groups of people, who thus stand in a privileged position in relation to that knowledge, those individuals or groups have a particular claim to access to that information.

Subject-producer reasons to share knowledge may be premised on consideration of those from whom the knowledge is derived – the data subjects may be thought to have an especial claim to sharing the knowledge, additional to the claim that it is "about" them. Complications may arise where generalisable knowledge is obtained from data produced from certain individuals as representatives of a group or of humanity as a whole, but where others could equally well have provided this data to produce the findings.

Material-producer reasons to share knowledge with those who provided economic or material support to its creation of knowledge and may apply to many different groups, including the researchers, the funders, and the subjects; on such grounds, private commercial entities may claim exclusive rights to knowledge that they produced and funded.

Beneficence reasons to share relate to claims that certain individuals or groups may benefit from the application of that knowledge.

Stewardship or custodian reasons to share knowledge may be premised upon particular duties of those in possession of certain collections of knowledge or information. There can be competing notions of in what good stewardship of data might consist.

10.8 Conclusions

Sharing data from genomics research has obvious great benefits for the advancement of science and therefore for the potential benefits for humanity as a whole that might result from this. A simple account of the ethical questions involved might pit ethics against science; or another simple account might pit privacy interests of data subjects with the interests of researchers in having access to resources. However there are many more issues involved, many parties who stand in different relations to the developing practices of data sharing, and different ways in which the parties involved might gain or lose from different data sharing policies and practices. There is a need to consider global issues whilst not losing sight of the need for local solutions and the importance of diversity within science and the scientific community. The issues are also heavily shaped by continually changing possibilities in science, technology and data manipulation. The complexity of these interrelated issues is such that unintended consequences could easily arise. Careful analysis of the practical and normative issues is essential.

10.9 Study Questions

1. (a) Would you give consent for your genomics data to be sequenced as part of a research project and shared freely with any researchers?
   (b) Would you give consent to your genomics data to be sequenced and posted freely on the web?
   Explain the reasons for your answer.

2. (a) Would you give consent for your genomics data to be shared with private pharmaceutical companies who stand to make a profit from any drugs that are eventually produced using knowledge derived from the data? Explain the reasons for your answer.
(b) A family member has been diagnosed with a genetic condition. A major pharmaceutical company is researching a new drug tailored to this condition. Would you give consent for this company to use your genomics information in researching this drug?
3. A group of researchers from Institution A have access to DNA samples and data. The consent forms that recruits signed made it explicit that the samples and data were only being used by Institution A. Institution B would like to use these samples in their research, but are refused permission. In response, Institution B turns down the request of Institution A to have access to samples they collected under blanket consent, where permission was given for the use in any medical research. Discuss the issues involved.
4. You volunteer for a genomics research project and sign consent for your data and sample to be used in any legitimate medical research and to be shared freely. The consent form stipulates that there will be no commercial involvement in the project and that the researchers will not gain personally from the use of your sample or data. The group maintains exclusive rights to publish material derived from the data for twelve months to safeguard their authorship rights. Discuss.
5. There have been suggestions that data should be made available only to "*bone fide*" researchers. Is this a helpful suggestion? How should the "*bone fide*" researcher be characterised?
6. Who makes a "good data sharer"?
7. In what does good stewardship of data consist?
8. For efficient data sharing to occur, there have been calls to harmonise data collection and to uphold high standards of data curation. Discuss any benefits, challenges or pitfalls of implementing such harmonisation.

References

Arribas-Ayllon M, Featherstone K, Atkinson P (2011) The practical ethics of genetic responsibility: non-disclosure and the autonomy of affect. Soc Theor Health 9:3–23

Biotechnology and Biomedical Sciences Research Council BBSCR data sharing policy. http://www.bbsrc.ac.uk/web/FILES/Policies/data-sharing-policy.pdf

Cancer Bioinformatics Grid Data sharing information. https://cabig.nci.nih.gov/

Church G, Heeney C, Hawkins N, de Vries J, Boddington P, Kaye J, Bobrow M, Weir B (2009) Public access to genome-wide data: five views on balancing research with privacy and protection. PLoS Genet 5(10):e1000665

Davies K (2001) Sequence: inside the race for the human genome. Phoenix, London

Delamothe T (1996) Whose data are they anyway? BMJ 312(7041):1241–1242

Featherstone K, Atkinson P, Bharadwaj A, Clarke A (2006) Risky relations: family, kinship and the new genetics. Berg, Oxford

HapMap Consortium SNP consortium. http://snp.cshl.org/. Accessed July 2011

International Strategy Meeting on Human Genome Sequencing (1997, 1996) Polices on releease of human genome sequence data. http://www.ornl.gov/sci/techresources/Human_Genome/research/bermuda.shtml

Kaye J, Heeney C, Hawkins N, de Vries J, Boddington P (2009) Data sharing in genomics: re-shaping scientific practice. Nat Rev Genet 10(5):331–335

Locke J (1689) Two treatises of government. Cambridge University Press, Cambridge

Medical Research Council Data sharing initiative. http://www.mrc.ac.uk/Ourresearch/Ethicsresearchguidance/Datasharinginitiative/index.htm

Moore AD (2000) Owning genetic information and gene enhancement techniques: why privacy and property rights may undermine social control of the human genome. Bioethics 14(2): 97–119

Moore v Regents of the University of California (1990). 793 P 2d 479 (Cal SC 1990)

National Institutes of Health Data sharing policy. http://grants2.nih.gov/grants/policy/data_sharing/

Organisation for Economic Co-operation and Development OECD guidelines for access to research data from public funding. http://www.oecd.org/dataoecd/9/61/38500813.pdf

Personal Genomes Project. www.personalgenomes.org. Accessed July 2011

Plato (1956) Protagoras and Meno (trans: Guthrie WKC). Penguin, Harmondsworth

Pope A (1796) An essay on man. Cadell and Davies, London

Redner H (1987) Pathologies of science. Soc Epistemol 1(3):215–247

Resnik DB (2007) The price of truth: how money affects the norms of science. Oxford University Press, Oxford

Skloot R (2010) The immortal life of Henrietta Lacks. Macmillan, London

Spinello RA (2004) Property rights in genetic information. Ethics Inform Technol 6:29–42

The Wellcome Trust Policy on data management and sharing. The wellcome trust. http://www.wellcome.ac.uk/About-us/Policy/Policy-and-position-statements/WTX035043.htm

The Wellcome Trust (2003) Sharing data from large-scale biological research projects: a system of tripartite responsibility. http://www.genome.gov/Pages/Research/WellcomeReport0303.pdf

Toronto International Data Release Authors (2009) Prepublication data sharing. Nature 461 (7261):168–170. doi:http://www.nature.com/nature/journal/v461/n7261/suppinfo/461168a_S1.html

Vickers AJ (2006) Whose data set is it anyway? Sharing raw data from randomised trials. Trials 7

Witte JId, Have Ht (1997) Ownership of genetic material and information. Soc Sci Med 45 (1):51–60

Afterword

At the start of this book, I tried to make it clear that it would not be possible to cover all the ethical issues that arise in relation to genomics research. As I reach the end of the book, it is even more apparent to me how many difficult, diverse, and finely nuanced questions have only been touched upon, and how many more have not even been raised. For example, I have merely indicated the broad sweep of questions that can arise about genomics research in the context of public health genomics, and have not raised many questions that arise from the application of genomics knowledge. I have mentioned biobanks, but have not discussed in detail the complex web of ethical questions that can be asked about them. I have tried to restrict the questions examined to those most immediately arising from the conduct of genomics research itself. Even here, I am painfully aware that for each question examined, there are layers of detail and debate that have had to be condensed and simplified. I am also of course aware that my introductions to reasoning and argument in ethics, and to many philosophical and ethical issues, are themselves very brief; a great deal more could be said on all these areas.

This shortcoming is partly a matter of pragmatics, in fitting a great many topics into the length of a book, and partly a matter of the nature of the field, where so many issues intersect, and where there are constant developments and surprises.

What I have rather aimed to do is to start a dissection of some of the major aspects of some central debates in the ethics of genomics research with the aim of encouraging and enabling further debate by others and fostering interdisciplinary understanding and dialogue. As I have argued, historically major abuses in research ethics and in genetics have been linked to a lack of such open dialogue and exchange of ideas. I have also dwelt upon the rift that so often occurs between ethical words and ethical action. If even just one reader has found anything in this book helpful in understanding and furthering discussion on the ethical challenges that we face in relation to genomics research, I will be happy. If this contributes in any way to furthering good practice, I will be happier still.

Oxford, August 2011 Paula Boddington

DOI 10.1007/978-3-642-23699-0,

Glossary

apoE A gene associated with risk of Alzheimer's disease.

Aristotle Greek philosopher, student of the philosopher Plato and teacher of Alexander the Great. He studied a wide range of subjects, including science and political philosophy. His best-known work in ethics is the *Nicomachean Ethics*.

Autonomy Literally, self-rule. See Chap. 6.

Beneficence The aim to produce good for another person or persons.

Bentham, Jeremy Early utilitarian philosopher and godfather to John Stuart Mill. Considered that we were governed by two "sovereign masters" pleasure and pain, and that the only relevant moral question to ask was not, "can it think, nor, can it reason, but can it suffer?"

Biobank Any repository of research samples and/or data for use in medical research, usually referring to large collections from many thousands of individuals.

Blanket consent Sometimes also called open or broad consent. In research, consent is given to future uses of one's data with only a rough idea of what these uses might be.

Capacity In ethics and in law, to have capacity is to have the ability to exercise valid choice. This must involve the ability to understand relevant information, to weigh it up, and to make a decision without undue outside influence.

Carrier In genetics, a person who does not have the symptoms of a genetic condition, but who has the potential to pass it on to an offspring.

Chromosome A strand of DNA found in the nuclei of cells. Humans have 46 paired chromosomes.

Cognitive dissonance A tension between different beliefs an individual has. See Chap. 3.

Cohort study A type of research design in which a group sharing certain characteristics is followed over a period of time, for example, a birth cohort study examines a group of people born in a certain time period over a certain geographical area.

P. Boddington, *Ethical Challenges in Genomics Research*,
DOI 10.1007/978-3-642-23699-0, © Springer-Verlag Berlin Heidelberg 2012

Confidentiality Some relationships give rise to duties of confidentiality, where one party (say, a doctor) has a duty not to disclose to third parties information imparted by the other party (say, a patient).

Consent In consenting to something, one is actively giving permission for it.

Consequentialism In ethics, the view that the only thing of relevance in ethics is the outcomes of actions. This contrasts with views that holds that features of agents, such as their motives or character traits, or features of actions, such as the rule they fall under, are what matters morally.

Control In research on a certain condition or trait, a control group of those who do not have this condition or trait is used to determine what characteristics are distinctive of those with the condition or trait of interest. Control groups are recruited from a group with similar characteristics (e.g., age, sex, and population group) in order to avoid confounding factors from affecting results.

Difference principle A method of ensuring fairness in distribution of goods across a society. See Sect. 6.2.7.

Dilemma, moral A genuine moral dilemma is a situation where there are incommensurable moral values involved, such that whatever you do, there is some moral cost. Moral philosophers disagree on whether there are any genuine moral dilemmas or if there are just situations in which we do not know what the best thing to do is.

Discourse analysis A discipline or approach to language which uses various techniques to interrogate and uncover features of language and communication.

Enlightenment An intellectual movement of the 18th century which stressed the importance of reason rather than tradition and which fostered scientific discovery

Epistemology The theory of knowledge, a branch of philosophy asking general questions about the justification of our beliefs.

Eudaimonia A Greek word often translated as "happiness" or "the good life".

Eugenics An attempt to "improve" the genetic stock of the human population. See Sect. 2.2.1.

Exploitation To exploit someone is to manipulatively use them for your own ends. See text box Sect. 9.5.

Framing Any means by which selective attention to any part of a text is achieved. See Sect. 4.5.

Genetic determinism A claim about causation that genes alone dictate biology.

Genetic essentialism A claim about who we are, that our genes give us the essence or core of our nature.

Genetic exceptionalism The claim that genetics has special properties that other areas of biology or of medicine do not possess.

Genetic reductionism A claim about explanation that we can explain other phenomena (such as other areas of biology or human behaviour) in terms solely of genetics.

Geneticisation The claim that phenomena which would be better explained in a variety of ways are explained more narrowly through genetics.

Genome-wide association study A research method in genomics where the entire genome is scanned for links with the condition or trait of interest. This can enable researchers to discover correlations without any prior hypothesis about where in the genome to look.

Genotype The complete genome of an organism.

Haemophilia A genetic condition where blood clotting factors are absent or reduced. Inherited via the X chromosome.

Hare, R.M. A utilitarian moral philosopher who held "universal prescriptivism", and considered that moral terms are to be taken to be prescriptions or commands that apply universally to all relevantly similar situations. Works include *The Language of Morals* and *Moral Thinking*.

Human nature A broad characterisation of the essential nature of human beings. There are many different philosophical theories about human nature and its relevance to political and moral questions. Many accounts attempt to spell out what is distinctive or special about human beings. Accounts vary on many factors, for example, on how fixed or malleable human nature is thought to be, the place of reason in human nature, and the social nature of humans.

Huntington's disease A genetic disease involving progressive neurological deterioration, invariably fatal, and for which there is currently no effective treatment.

Instrumental Something is valued instrumentally if it is valued as a means to something else.

Intrinsic Something is valued intrinsically if it is valued for its own sake, quite apart from what instrumental value it might (additionally) have.

Longitudinal study A research study that follows recruits over a long period of time, for example, to study development or disease pathogenesis or risk factors.

Metaethics The study of questions about the nature of ethics, such as whether there are any universal moral facts, how we can have moral knowledge, and how disagreements in ethics are to be solved.

Mill, John Stuart A utilitarian philosopher and social reformer, briefly MP for Westminster, author of *Utilitarianism*, *On Liberty*, and *The Subjection of Women* as well as many other works.

Mitochondrial DNA Small fragments of DNA present in the mitochondria (the energy centres) of cells. Unlike chromosomal DNA, this DNA is passed down the maternal line (to both sons and daughters).

Non-maleficence The duty of not causing harm to others.

Paternalism Literally, treating someone as a father might treat a child; acting on the assumption that you know what is in someone else's best interests.

Pharmacogenomics The application of genomics knowledge to pharmaceuticals to tailor drugs to individuals to ascertain best dosage, avoid side effects, and determine the most effective drug for each individual.

Phenotype The biological nature of an organism.

Phronesis A Greek word often translated as practical wisdom, referring to the development of ethical judgement. See text box Sect. 1.4.3.

Pre-implantation genetic diagnosis Techniques to ascertain if an embryo (created in vitro) has certain genetic traits before implantation to the mother.

Pre-natal diagnosis Techniques used to determine if a fetus in utero has certain genetic or other medical conditions.

Quantitative trait A biological trait that varies continuously between individuals such as height or blood pressure.

Rational agent A term often used in philosophy to refer to any being that possesses the capacity to reason. Some human beings may not be rational agents (e.g., babies, the comatose) and some rational agents may not be humans (e.g., angels).

Rawls, John Widely held to be the foremost political philosopher of the 20th century, author of *A Theory of Justice*. See also text box on Reflective equilibrium, Sect. 3.2.6.

Relativism In ethics, descriptive relativism observes that there are different moral codes applicable in different societies and at different times: normative relativism holds that we can only make moral judgements relative to a particular society or group.

Rights To have a right is to have a claim against someone else in relation to that right: either negatively, not to prevent one from acting in a certain way, or positively, to provide certain goods one has a right to. Some hold that there are natural rights, which all have by virtue of being born a human being, others that there are only positive rights, as laid down in particular legal systems.

SNP Single nucleotide polymorphism – a single point of variation in the genome of an individual, often the subject of genomics research into variation between individuals with respect to diseases or quantitative traits.

Stigma Social disapproval directed at a person group or trait. See text box Sect. 5.2.10.

Supererogation Actions over and above the call of duty.

Universalism in ethics The idea that ethical judgements must apply to all relevantly similar situations.

Utilitarianism A group of related ethical theories which hold that moral actions should be judged solely in terms of their tendency to produce happiness, or avoid unhappiness, for the greatest number of people.

Wittgenstein, Ludwig Austrian twentieth-century philosopher renowned for his characteristic approach to philosophy – author of *Tractatus Logico-Philosophicus* and *Philosophical Investigations*.

X chromosome One of the sex chromosomes. Females have two copies, males have one.

Y chromosome One of the sex chromosomes. Males have one Y chromosome, females none. It is passed down virtually unchanged from father to son.

Abbreviations

CEU	Utah residents with Northern and European Ancestry: a HapMap sample.
GWAS	Genome-wide association study
HapMap	The International Haplotype Map Project
HGC	Human Genetics Commission
HGP	Human Genome Project
PCR	Polymerase chain reaction
IRB	Institutional Review Board
NIH	National Institutes of Health (USA)
ELSI	Ethical, Legal and Social Implications
REC	Research Ethics Committee
SNP	Single nucleotide polymorphism

P. Boddington, *Ethical Challenges in Genomics Research*,
DOI 10.1007/978-3-642-23699-0,

Index of Text Boxes

P. Boddington, *Ethical Challenges in Genomics Research*,
DOI 10.1007/978-3-642-23699-0, © Springer-Verlag Berlin Heidelberg 2012

Subject Index

DOI 10.1007/978-3-642-23699-0,

MIX
Papier aus verantwortungsvollen Quellen
Paper from responsible sources
FSC® C105338

If you have any concerns about our products,
you can contact us on
ProductSafety@springernature.com

In case Publisher is established outside the EU,
the EU authorized representative is:
Springer Nature Customer Service Center GmbH
Europaplatz 3, 69115 Heidelberg, Germany

Printed by Libri Plureos GmbH
in Hamburg, Germany